SpringerBriefs in Mathen

M000268206

SpringerBriefs in Mathematics showcase expositions in all areas of mathematics and applied mathematics. Manuscripts presenting new results or a single new result in a classical field, new field, or an emerging topic, applications, or bridges between new results and already published works, are encouraged. The series is intended for mathematicians and applied mathematicians.

For further volumes:
http://www.springer.com/series/10030

Peter Buchholz • Jan Kriege • Iryna Felko

Input Modeling with Phase-Type Distributions and Markov Models

Theory and Applications

Springer

Peter Buchholz
Department of Computer Science
Technical University of Dortmund
Dortmund, Germany

Jan Kriege
Department of Computer Science
Technical University of Dortmund
Dortmund, Germany

Iryna Felko
Department of Computer Science
Technical University of Dortmund
Dortmund, Germany

ISSN 2191-8198 ISSN 2191-8201 (electronic)
ISBN 978-3-319-06673-8 ISBN 978-3-319-06674-5 (eBook)
DOI 10.1007/978-3-319-06674-5
Springer Cham Heidelberg New York Dordrecht London

Library of Congress Control Number: 2014939378

Mathematics Subject Classification (2010): 60J22, 60J28, 62M05, 65C40, 68M20, 68U20, 60K25, 65C10, 65C40, 90B15, 90B18, 90B25

Springer is part of Springer Science+Business Media (www.springer.com)

Preface

Nowadays system analysis of man-made systems, like computer systems, communication networks, and manufacturing plants, and also of natural systems, like biological or social systems, is often model based. To capture the complexity of real systems stochastic discrete event models are used in many application areas. One of the key aspects in building such models is the adequate description of real processes and event streams in a stochastic model. Often simple distributions are not sufficient for this purpose because observed distributions are multimodal and events are correlated.

One class of stochastic models, which allows one to describe multimodal distributions and correlated event times, are Markov processes with marked transitions. Since Markov processes can be analyzed with numerical methods and with stochastic simulation, they are an ideal candidate to describe event times in stochastic models. However, the big disadvantage of using Markov processes instead of simple distributions or stochastic processes, like autoregressive or moving average time series, is the parameterization effort. Usually, the finding of adequate parameters of a Markov model, to capture some observed behavior, is a non-linear optimization problem with many parameters and non-unique representations of a given stochastic distribution or process. This often prohibits the wider use of those models, in particular in stochastic simulation, and is the reason that only fairly simple phase-type distributions can be found in textbooks on stochastic modeling or simulation.

This books summarizes our work on the parameterization of phase-type distributions and Markovian arrival processes, which are the commonly used model types in modeling event streams. To the best of our knowledge it is the first time that the available methods are collected in a textbook. We hope that this helps to support the use of the mentioned models in stochastic modeling and in particular in stochastic simulation.

We thank our colleagues Axel Thümmler, Andriy Panchenko, and Falko Bause who worked together with us on several aspects of parameter fitting for Markovian arrival processes. Peter Kemper from the college of William and Marry (Williamsburg, VA, USA) spent his sabbatical in our group and worked together with us on the parameterization of Markovian arrival processes with multiple event times. We learned a lot about phase-type distributions and Markovian arrival processes from Miklós Telek, Gábor Horváth, and Levente Bodrog from the Stochastic Modelling Laboratory of the Department of Telecommunications, Technical University Budapest. We thank them for a long lasting and very fruitful cooperation. The research work has been supported by the *Deutsche Forschungsgemeinschaft* and by the *Deutsche Akademische Austauschdienst*.

Dortmund, Germany — Peter Buchholz
Dortmund, Germany — Iryna Felko
Dortmund, Germany — Jan Kriege
March 2014

Contents

Acronyms

ALS	Alternating least squares approach
APHD	Acyclic phase-type distribution
BMAP	Batch Markovian arrival process
BS	Basic series of an APH
cdf	Cumulative distribution function
CTMC	Continuous time Markov chain
EM	Expectation maximization
HErD	Hyper-Erlang distribution
IPP	Interrupted Poisson process
KPC	Kronecker product composition
MAP	Markovian arrival process
ME	Matrix exponential
ML	Maximum-likelihood
MMAP	Marked Markovian arrival process
MMPP	Markov modulated Poisson process
MRAP	Marked rational arrival process
NNLS	Non-negative least squares approach
PH	Phase-type
PHD	Phase-type distribution
QBD	Quasi-birth-death process
QN	Queueing network
RAP	Rational arrival process

Notation

$\mathbf{0}$	Matrix or vector where every entry is 0
$\mathbb{1}$	(Column) vector where every entry is 1
$APHD(n)$	APHD of order n
$\mathbb{B}$	Set of boolean values
B_i	Number of times a PHD or MAP starts in phase i
C, C^2	(Squared) coefficient of variation
$\mathbf{D}_0$	Matrix of internal transition rates of a PHD or MAP
$\mathbf{D}_1$	Matrix of transition rates generating an event of a MAP
$\mathbf{d}_1$	Exit vector of a PHD
$E[X]$	Expectation of random variable X
$\mathbf{H}_n$	Matrix of factorial moments up to order n
$\mathbf{I}$	Identity matrix
$\lambda(i)$	Transition rate out of state i
$\lambda(i, j)$	Transition rate between the states i and j
$\mathcal{L}(\mathbf{\Theta})$	Likelihood function
$\mathbf{M}$	$= -\mathbf{D}_0^{-1}$ moment matrix of a PHD or MAP, i.e. fundamental matrix of an absorbing Markov chain
$MAP(n)$	MAP of order n
μ_i	i-th moment
$\hat{\mu}_i$	Estimator of the i-th moment
μ_{kl}	Joint moment of order k, l of two consecutive inter-event times
$\hat{\mu}_{kl}$	Estimator of the joint moment of order k, l of two consecutive inter-event times
M_{ij}	Number of transitions from phase i to j with generating an event
n	Order of a PHD or MAP
n_i	i-th normalized moment
N_{ij}	Number of transitions from phase i to j without generating an event
$N(t)$	Number of events of a counting process in the interval $[0, t]$
$\mathbb{N}$	The set of natural numbers
Ω, Ω_I	Partition of the state space and partition group
$\boldsymbol{\pi}$	Initial vector of a PHD or a MAP

$\boldsymbol{\pi}_s$	Stationary vector of a MAP at event generation time points
p_t	Probability distribution of a CTMC at time t
$\mathbf{P}$	Transition probability matrix of an embedded Markov process of a continuous-time Markov-chain
$\mathbf{P}_0$	$= \mathbf{D}_0/\alpha + \mathbf{I}$, matrix of the discrete time Markov-chain used for uniformization
$\mathbf{P}_1$	$= \mathbf{D}_1/\alpha$, matrix of transitions related to events in the discrete time Markov-chain used for uniformization
$\mathbf{P}_s$	$= (-\mathbf{D}_0)^{-1}\mathbf{D}_1$, matrix of the discrete time Markov-chain at event generation time points
$PHD(n)$	PHD of order n
$\mathcal{P}$	Process (describing a real system or an adequate simulation model)
$\mathbf{Q}$	Infinitesimal generator matrix of a continuous-time Markov-chain
ρ_k	Coefficient of autocorrelation at lag k
$\hat{\rho}_k$	Estimator for the coefficient of autocorrelation at lag k
r_i	i-th factorial moment
$\mathbb{R}$	Set of real numbers
$\mathcal{S}$	State space of a CTMC
$\mathcal{S}_T$	Set of transient states of an absorbing CTMC
$\mathcal{S}_A$	Set of absorbing states of an absorbing CTMC
$\mathcal{T}$	Trace, i.e. a sequence of observations usually measured from a real system
$\mathcal{T}^*, \bar{\mathcal{T}}$	Aggregated trace
$\tilde{\mathcal{T}}$	Grouped trace
$\mathbf{V}$	Collector matrix of an aggregation
$VAR[Y]$	Variance of random variable Y
$\mathbf{W}$	Distributor matrix of an aggregation
$X(t)$	Stochastic process
Z_i	Total time spent in phase i of a PHD or MAP before generating an event

Chapter 1
Introduction

Quantitative analysis of man-made systems like computer systems, communication networks, manufacturing plants, logistics networks, to mention only few examples, is often done by means of discrete event models that are analyzed numerically [152] or by simulation [105]. One key issue in these models is the adequate modeling of the load which describes the occurrence of events, let it be customer arrivals in queueing networks, failure times in reliability models or packet lengths in simulation models of computer networks. In more abstract terms one can think of arrival, service or failure times that are part of a model. We will use the term *inter-event times* to capture the different quantities in a model. Inter-event times are characterized by random variables or stochastic processes generating non-negative numbers.

The choice of an adequate model for the load of a system often depends on the used model type. In simulation, modeling of the load is subsumed under the term *input modeling* [14, 105]. Usually, this means that a stochastic model is generated to capture the key features of an input process for which some measurements or estimates are available. In the past, often independently identically distributed events are assumed and one distribution from a set of available distributions is selected to describe the observed input. For this purpose software tools are available that perform an automatic parameter fitting, the selection of an adequate distribution and an integration of the generated distribution in a simulation program [92, 106]. However, often the set of available distributions is not flexible enough to capture measured behavior or the assumption of independent and identically distributed events does not hold because events are correlated as it is usually the case for the load in computer networks [109, 140] or sometimes the case with failures in complex systems [63,128,141]. In such situations, simulation literature recommends the use of empirical distributions, time series models or multivariate normal or Johnson distributions [14,105]. However, support for these approaches is usually not available such that stochastic models have to be built and integrated in a simulation model manually which is cumbersome and error-prone.

P. Buchholz et al., *Input Modeling with Phase-Type Distributions and Markov Models: Theory and Applications*, SpringerBriefs in Mathematics,
DOI 10.1007/978-3-319-06674-5_1,

If models should be analyzed numerically or analytically, usually Markov processes are the underlying stochastic processes. This implies that events have to be triggered by Markov processes. Resulting stochastic models are phase-type distributions (PHDs) to describe service or arrival times in queueing networks [125, 148]. However, although it is known that PHDs are very flexible and allow one to approximate general distributions on the positive axis arbitrarily close [132], the use of PHDs was in the past mainly restricted to a few subclasses, like Erlang- or hyper-exponential-distributions which are parameterized according to the first two moments of the measured inter-event times [148]. The approximation of a general distribution by a PHD is a complex non-linear optimization problem for which only recently computational algorithms have been proposed which have not found their way into broadly available modeling software yet. Markov process based modeling can also be applied to describe correlated inter-event times by using *Markovian Arrival Processes* (MAPs) [124]. The analysis of single queues with MAP input or even service is established and is based on well known matrix analytic techniques [104, 125]. However, parameter fitting for MAPs is even more complex than parameter fitting of PHDs [76]. Recently, several algorithms for generating MAPs from measured data became available [25, 31, 32, 41, 58, 82, 94, 97, 133, 155] but these approaches are not widely established in stochastic modeling yet. MAPs, like PHDs, can also be integrated in simulation models but again this approach is not really supported by available simulation tools because first approaches describing the integration of MAPs in simulation models have been published only recently [11, 27, 68].

PHDs and MAPs are very flexible stochastic models which allow one, if adequately parameterized, to capture a variety of behaviors as they are required in stochastic modeling of discrete event systems. In contrast to other distributions or stochastic processes, the range of behaviors that can be expressed is extremely wide such that, at least theoretically, a single model type, namely Markov processes, is sufficient for modeling all processes in a model. However, the price for the flexibility is the huge effort to find the parameters such that the resulting model approximates the observed or required behavior sufficiently close. The complexity of parameter estimation approaches prohibited in the past a wider use of PHDs and MAPs in applied stochastic modeling. The current situation is characterized by a large number of papers, mainly in queueing theory, how to solve models including PHDs or MAPs and many theoretical papers describing features of PHDs and MAPs are available. Additionally, several papers on parameter fitting for PHDs and much fewer papers on the parameter fitting of MAPs exist. All this material is available in conference and journal publication, there are almost no textbooks on applied modeling that consider PHDs or MAPs in a broader context. If PHDs are introduced, the description is often restricted to Erlang- or hyper-exponential distributions and the simple fitting of the first two moments using these distributions.

Our feeling is that the mentioned situation is unsatisfactory and does not reflect the state of the art. Although parameterization of PHDs and MAPs is still a challenge, a wide variety of techniques are ready to be used in practice and allow one to generate an adequate model from measured data or from some characteristics derived from a real system. The intention of this book is to give an application oriented introduction of PHDs and MAPs, to present available methods for parameterization of these models and to show how the resulting Markov models can be integrated in quantitative models like queueing networks or simulation models. In Chap. 2 PHDs are introduced, specific subclasses that allow an efficient parameterization are presented and equivalences are defined among different representations of the same distribution. Chapter 3 summarizes available algorithms for parameter fitting of PHDs and presents in detail some algorithms that have shown to work in a wider context. In the Chaps. 4 and 5 the same information is given for MAPs which can be seen as an extension of PHDs. After the basic steps for generating PHDs and MAPs have been introduced, the following two chapters are devoted to practical aspects. The embedding of PHDs and MAPs in different models and the introduction of some application examples are treated in Chap. 6. In Chap. 7, available software for generating PHDs, MAPs or for analyzing models including these processes is introduced. The book ends with conclusions which give an outlook of current research questions in input modeling using Markov models.

The book considers input modeling using *Continuous Time Markov Chains* (CTMCs) with marked transitions which can be seen as a superclass of PHDs and MAPs. It is possible and in some applications adequate, to use a discrete rather than continuous time scale resulting in discrete time models. It is indeed possible to define PHDs [20, 48, 147] and MAPs [1, 158] in discrete time and several results can be easily transferred from the continuous to the discrete time scale. We do not consider discrete time models in this book to limit the overall length of the book. Other classes of processes which are strongly related to PHDs and MAPs are Matrix Exponential (ME) distributions [111] and Rational Arrival Processes (RAPs) [5]. These processes are defined purely algebraically such that the intuitive stochastic interpretation is lost. Although it has been shown recently that these general models and Markov models are strongly related [37, 38], the use of ME distributions and RAPs in practical modeling is currently very limited. Therefore we decided not to consider them in the book.

Chapter 2
Phase-Type Distributions

2.1 Basic Definitions

Continuous-time Markov chains (CTMCs) are a class of stochastic processes with a discrete state space in which the time between transitions follows an exponential distribution. In this section, we first provide the basic definitions for CTMCs and notations associated with this model. We then proceed with an explanation of the basic concepts for phase-type distributions (PHDs) and the analysis of such models. For theoretical details about CTMCs and related stochastic processes we refer to the literature [151].

2.1.1 *Markov Chains*

Let $\mathcal{S}$ denote a countable set of states, and let $\{X(t)\}_{t\geq 0}^{\infty}$ be a stochastic process with state space $\mathcal{S}$. Let $n \in \mathbb{N} \cup \{\infty\}$ be the size of the state space. Since the state space is isomorphic to (a subset of) $\mathbb{N}$, we denote states by their numbers.

$\{X(t)\}_{t\geq 0}^{\infty}$ is a continuous-time Markov chain, if it is characterized by the *Markov property* [151]:

$$Prob(X(t_{k+1}) = x_{k+1}|X(t_k) = x_k, \ldots, X(t_0) = x_0) = Prob(X(t_{k+1}) = x_{k+1}|X(t_k) = x_k), \tag{2.1}$$

for any $0 \leq t_0 \leq t_1 \leq \ldots \leq t_k \leq t_{k+1}$ and $x_l \in \mathcal{S}$. Thus, indicating that given the current state x_k and the time t_k, the value taken by $X(t_{k+1})$ depends only on x_k and on t_k but not on the past of the process.

The process is time homogeneous if for all $u \geq 0$

$$Prob(X(t+u) = j|X(u) = i) = Prob(X(t) = j|X(0) = i) = p_t(i,j). \tag{2.2}$$

P. Buchholz et al., *Input Modeling with Phase-Type Distributions and Markov Models: Theory and Applications*, SpringerBriefs in Mathematics,
DOI 10.1007/978-3-319-06674-5_2,

The *homogeneity* in the definition of homogeneous Markov chains is due to the fact that the transition probabilities only depend on the difference t between u and $t+u$ and not on the actual times $(u,t+u)$. The values $p_t(i,j)$ define a matrix with transition probabilities $\mathbf{P}_t$.

The state probabilities at time t are denoted by $p_t(j) = Prob(X(t) = j),\ j \in \mathcal{S}$ with $\sum_j p_t(j) = 1$. Consequently, the vector $\boldsymbol{\pi}(0) = [p_0(1), p_0(2), \ldots]$ is the initial probability vector of the CTMC. Moreover, the random times between state transitions are exponentially distributed random variables $V_1, V_2, \ldots$ with parameter $\lambda(i)$ for state i. That is, each V_i describes the exponential holding time in state i and has distribution function

$$Prob(V_i \leq t) = 1 - e^{-\lambda(i)t}, \quad t \geq 0. \tag{2.3}$$

Such a process evolves as follows: at any time t, $X(t) = i$, the process remains in state i for a period of time determined by an exponential distribution with parameter $\lambda(i)$, $0 \leq \lambda(i) < \infty$, and then jumps to a state j with probability $p(i,j) = \lambda(i,j)/\lambda(i)$. Therefore, $\lambda(i,j)$ is the rate at which a state transition from i to j occurs. Moreover, $\lambda(i)$ represents the total event rate characterizing state i. We now summarize the probabilistic behavior of the continuous time Markov chain in terms of its infinitesimal generator [151]. The infinitesimal generator is a $n \times n$ matrix $\mathbf{Q}$ with components

$$\mathbf{Q}(i,j) = \begin{cases} -\lambda(i) & \text{if } i = j, \\ \lambda(i,j) & \text{if } i \neq j. \end{cases} \tag{2.4}$$

$\mathbf{Q}$ is also denoted as the transition rate or generator matrix. Since $\lambda(i) \geq 0$, it follows $\mathbf{Q}(i,i) \leq 0$ indicating that all diagonal elements of the matrix $\mathbf{Q}$ are non-positive. If the transition to some state j is feasible in state i then $\mathbf{Q}(i,j) > 0$, otherwise, $\mathbf{Q}(i,j) = 0$. Thus, all non-diagonal elements must be non-negative. One can now see, that from the definition of the rates it follows that

$$\sum_j \mathbf{Q}(i,j) = 0. \tag{2.5}$$

The embedded process $\{X_r\}_{r \in \mathbb{N}_0}$, with $X_0 = X(0)$, is a discrete-time Markov chain with transition probability matrix $\mathbf{P}$. It applies for the elements $\mathbf{P}(i,j) = Prob(X(r+1) = j | X(r) = i)$ which can be expressed in terms of $\mathbf{Q}$

$$\mathbf{P}(i,j) = \frac{\mathbf{Q}(i,j)}{-\mathbf{Q}(i,i)}, \quad \text{for } j \neq i,\ \mathbf{Q}(i,i) \neq 0, \tag{2.6}$$

and $\mathbf{P}(i,i) = 0$ in this case. For $\mathbf{Q}(i,i) = 0$ also $\mathbf{Q}(i,j) = 0$ for all $j \neq i$. In this case we define $\mathbf{P}(i,i) = 1$ and $\mathbf{P}(i,j) = 0$ for $i \neq j$. States with $\mathbf{Q}(i,i) = 0$ are denoted as absorbing states. Note that summing over all j results in $\sum_j \mathbf{P}(i,j) = 1$.

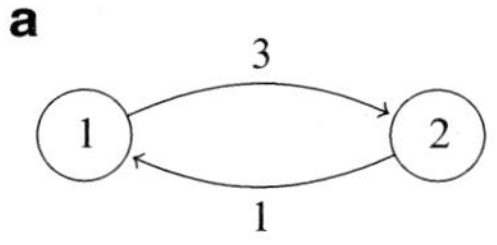

b $Q = \begin{bmatrix} -3 & 3 \\ 1 & -1 \end{bmatrix}$

Fig. 2.1 State transition diagram and generator matrix of a CTMC. (**a**) A state transition diagram of the CTMC. (**b**) The infinitesimal generator **Q**

Markov chains can be equivalently described by the generator matrix **Q** or by a state transition diagram as shown in Fig. 2.1. The edge connecting states i and j is weighted with the transition rate from i to j, i.e., with $\mathbf{Q}(i, j)$.

2.1.2 *Absorbing Markov Chains and Phase-Type Distributions*

Next we consider Markov chains with absorbing states. We will introduce several definitions and classifications of the states of a Markov chain in ways that turn out to be particularly convenient for our purposes, namely the study of the behavior of the chain up to the moment that it enters an absorbing state. The states of the Markov chain can be classified according to whether it is possible to move from a given state to another given state. Clearly, if a state j is reachable from a state i we have $p_t(i, j) = Prob(X(t+u) = j|X(u) = i) > 0$ for some t, i.e., the process can move from state i to state j after some amount of time t. A complete treatment of this classification can be found in [93, 151]. We will take a few concepts from that description.

Definition 2.1. States i and j can communicate with each other if i is reachable from j and vice versa.

Let $\mathcal{C}$ be a subset of the state space $\mathcal{S}$. If all states in set $\mathcal{C}$ communicate, we call it a communicating set. If there no feasible transition from any state in $\mathcal{C}$ to any state outside $\mathcal{C}$, then $\mathcal{C}$ forms a closed set:

Definition 2.2. A subset $\mathcal{C}$ of the state space $\mathcal{S}$ is said to be closed if $\mathbf{P}(i, j) = 0$ for any $i \in \mathcal{C}$, $j \notin \mathcal{C}$.

If $\mathcal{C}$ consists of a single state, say i, then i is said to be an absorbing state.

A closed set $\mathcal{C}$ where all members communicate is a closed communicating set.

Clearly, if i is an absorbing state it holds that $\mathbf{P}(i,i) = 1$. A process can never leave a closed set after entering it. The following two classifications indicate whether and when a process returns to a state after leaving it.

Definition 2.3. A state $i \in \mathcal{S}$ is a transient state, if the probability of returning to i after leaving it is less than 1.

A state $i \in \mathcal{S}$ is a recurrent state, if the probability of returning to i after leaving it is 1. If the mean time to return to i is finite, then the state is positive recurrent otherwise null recurrent.

All states in a finite closed communicating set are positive recurrent. Every state space $\mathcal{S}$ can be partitioned into closed communicating subsets $\mathcal{C}_I \subset \mathcal{S}$ $(I = 1,\ldots,N)$ and the remaining states collected in subset $\mathcal{O} = \mathcal{S} \setminus \{\cup_{I=1}^{N} \mathcal{C}_I\}$. If $\mathcal{O}$ is non-empty, then it contains states i that cannot be entered from a state in one of the closed communicating subsets $\mathcal{C}_I$ but there is a non-zero probability to enter at least one of the subsets $\mathcal{C}_I$ starting in i which implies that all states in $\mathcal{O}$ are transient. This also implies that in a CTMC with a finite state space every state is positive recurrent or transient. The situation is more complex for infinite state spaces because in a closed communicating set of an infinite size, states can be transient, null recurrent or positive recurrent.

If every state in a Markov chain is either absorbing or transient, then the Markov chain is called *absorbing Markov chain*. A particularly interesting case of absorbing Markov chains is one consisting of a single absorbing state. These chains will be central in this book.

We assume here that the state space $\mathcal{S}$ of the continuous time absorbing Markov process $\{X(t)\}_{t\geq 0}^{\infty}$ is finite and contains the set of transient states $\mathcal{S}_T = \{1,\ldots,n\}$ and a single absorbing state $n+1$. We order the states of the CTMC such that the n transient states occur first and can write the infinitesimal generator matrix $\mathbf{Q}$ as

$$\mathbf{Q} = \left[\begin{array}{c|c} \overbrace{\mathbf{D}_0}^{n} & \overbrace{\mathbf{d}_1}^{1} \\ \hline \mathbf{0} & 0 \end{array} \right] \begin{array}{l} \}\, n \\ \}\, 1 \end{array} \tag{2.7}$$

Combining all transient states together we obtain a $n \times n$ submatrix $\mathbf{D}_0$ describing only transitions between transient states. The $n \times 1$ vector $\mathbf{d}_1$ contains transition intensities from transient states to the absorbing state. The row vector $\mathbf{0}$ consists entirely of 0's since no transition from the absorbing state to transient states can occur. The remaining element of the matrix $\mathbf{Q}$ is 0 and gives the transition rate out off the absorbing state. Consider the absorbing chain with transition matrix $\mathbf{Q}$ in Fig. 2.2. State 4 is absorbing, hence the transition rates to other states are 0. The regions of the matrix $\mathbf{Q}$, namely $\mathbf{D}_0$, $\mathbf{d}_1$, and $\mathbf{0}$ are marked.

Since the states are transient, matrix $\mathbf{D}_0$ is nonsingular [103, Theorem 2.4.3], i.e. $\lim_{t\to\infty} Prob(X(t) < n+1) = 0$. The stated theorem establishes that the absorption occurs with probability 1. Furthermore, matrix $(-\mathbf{D}_0)^{-1}$ is the fundamental matrix of the absorbing continuous time Markov chain as defined in [93]. The value $(-\mathbf{D}_0)^{-1}(i,j)$ gives the expected total time spent in state j before absorption given that the initial state is i.

a

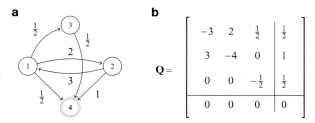

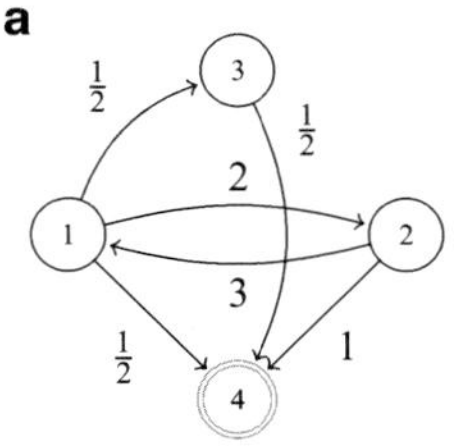

b

$$\mathbf{Q} = \left[\begin{array}{ccc|c} -3 & 2 & \frac{1}{2} & \frac{1}{2} \\ 3 & -4 & 0 & 1 \\ 0 & 0 & -\frac{1}{2} & \frac{1}{2} \\ \hline 0 & 0 & 0 & 0 \end{array}\right]$$

Fig. 2.2 An absorbing CTMC with absorbing state 4. Hence $\mathbf{D}_0$ is 3×3 and $\mathbf{d}_1$ is 3×1 in this example. (**a**) The state transition diagram for the absorbing CTMC. (**b**) The matrix for the absorbing CTMC

Now we are able to introduce the concept of the phase-type distribution (PHD) along the lines of Neuts [124].

Definition 2.4. A phase-type distribution (PHD) is defined as the distribution of the lifetime X, i.e., the time to enter an absorbing state from the set of transient states $\mathcal{S}_T$ of an absorbing continuous time Markov process $\{X(t)\}_{t\geq 0}^{\infty}$.

The finite state space $\mathcal{S}$ of the continuous time absorbing Markov process $\{X(t)\}_{t\geq 0}^{\infty}$ contains the set of transient states $\mathcal{S}_T = \{1,\ldots,n\}$ and the set of absorbing states $\mathcal{S}_A = \{n+1\}$. The transient states are called *phases*. A PHD with n transient states is said to have *order n*. Furthermore the background CTMC $\{X(t)\}_{t\geq 0}^{\infty}$ has an initial probability vector $[\boldsymbol{\pi}, \boldsymbol{\pi}(n+1)]$ and the infinitesimal generator $\mathbf{Q}$ given in Eq. (2.7). In particular, $\boldsymbol{\pi}$ is a row vector of size n, $\boldsymbol{\pi}(n+1)$ is the initial probability for the absorbing state $n+1$. Note that $\mathbf{Q}$ is the intensity matrix thus indicating that the rows sum to zero as shown in Eq. (2.5). Therefore,

$$\mathbf{D}_0\mathbf{1} + \mathbf{d}_1 = \mathbf{0}, \tag{2.8}$$

where $\mathbf{1}$ is the column n-vector of 1's and $\mathbf{0}$ is the column n-vector of 0's. The sum 2.8 in matrix notation can be rewritten as $\mathbf{d}_1 = -\mathbf{D}_0\mathbf{1}$. In particular, the matrix $\mathbf{D}_0$ is a subintensity and we have that

$$\mathbf{D}_0(i,i) \leq 0,\ \mathbf{D}_0(i,j) \geq 0 \text{ for } i \neq j,\ \mathbf{d}_1(i) \geq 0 \text{ and } \sum_{j\in\mathcal{S}_T} \mathbf{D}_0(i,j) \leq 0. \tag{2.9}$$

The Markov process starts in an arbitrary state from $\mathcal{S} = \mathcal{S}_T \cup \mathcal{S}_A$. The vector $\boldsymbol{\pi} = [\pi(1),\ldots,\pi(n)]$ describes the initial probabilities for transient states and $\pi(n+1)$ gives the probability for a direct start in the absorbing state, called point mass at zero. Thus, it holds that $\boldsymbol{\pi}\mathbf{1} + \pi(n+1) = 1$. In most cases it is assumed that $\pi(n+1) = 0$ and there is no start in the absorbing state $n+1$. Then, $\boldsymbol{\pi}\mathbf{1} = 1$, so that X is strictly positive, which we will assume in the following.

a

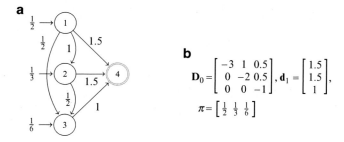

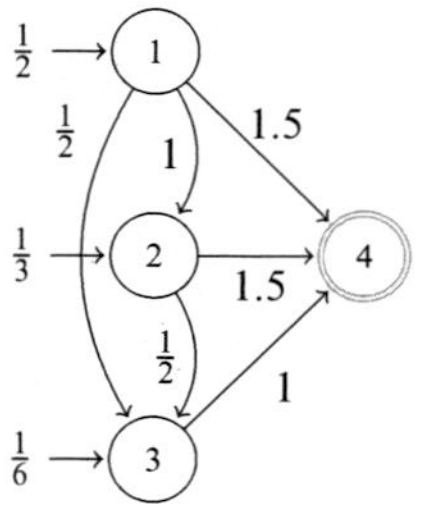

b

$$\mathbf{D}_0 = \begin{bmatrix} -3 & 1 & 0.5 \\ 0 & -2 & 0.5 \\ 0 & 0 & -1 \end{bmatrix}, \mathbf{d}_1 = \begin{bmatrix} 1.5 \\ 1.5 \\ 1 \end{bmatrix},$$

$$\pi = \begin{bmatrix} \frac{1}{2} & \frac{1}{3} & \frac{1}{6} \end{bmatrix}$$

Fig. 2.3 Symbolic representation of the PHD of Example 2.1. (**a**) A state transition diagram of a PHD. (**b**) The sub generator matrix $\mathbf{D}_0$, the exit-rate vector $\mathbf{d}_1$ and the initial probabilities $\boldsymbol{\pi}$ of the PHD

Since the background Markov process $\{X(t)\}_{t\geq 0}^{\infty}$ is a CTMC, the holding time of each phase i, $1 \leq i \leq n$, is exponentially distributed with parameter $-\mathbf{D}_0(i,i)$ as explained in Eq. (2.3). We obtain for diagonal elements of the subintensity $\mathbf{D}_0$

$$\mathbf{D}_0(i,i) = -\left(\sum_{j \neq i} \mathbf{D}_0(i,j) + \mathbf{d}_1(i) \right). \tag{2.10}$$

The column vector $\mathbf{d}_1$ has the interpretation of the exit rates, i.e. $\mathbf{d}_1(i)$ gives the intensity in state i for leaving $\mathcal{S}_T$ and going to the absorbing state $n+1$.

We now say that the random variable X describing the time till absorption is of phase-type with representation $(\boldsymbol{\pi}, \mathbf{D}_0)$. The vector $\mathbf{d}_1$ and the value $\pi(n+1)$ are implicitly given by the matrix $\mathbf{D}_0$ and vector $\boldsymbol{\pi}$, which is the reason for excluding them in the representation.

Example 2.1. Consider a PHD with the subgenerator $\mathbf{D}_0$ given in Fig. 2.3b. Here all states are entry states, since $\pi(i) \neq 0$ is satisfied for all transient states i. Furthermore all transient states i are exit states. This is the case if it is possible to reach an absorbing state directly from the state i.

2.1.3 *Analysis of Phase-Type Distributions*

We shall now give the basic analytic properties of PHDs. First of all we recall that the matrix exponential $e^{\mathbf{Q}}$ is defined by the standard series expansion $\sum_{k\geq 0} \frac{1}{k!}\mathbf{Q}^k$ (see, e.g. [103]). The distribution function of a phase-type distributed variable with representation $(\boldsymbol{\pi}, \mathbf{D}_0)$ is given by

$$F(x) = 1 - \boldsymbol{\pi} e^{\mathbf{D}_0 x}\mathbf{1} \text{ for } x \geq 0 \tag{2.11}$$

and its associated density function is given by

$$f(x) = \boldsymbol{\pi} e^{\mathbf{D}_0 x} \mathbf{d}_1 \text{ for } x \geq 0. \tag{2.12}$$

Let us assume that a Markov process $\{X(t)\}_{t\geq 0}^{\infty}$ with an infinitesimal generator $\mathbf{Q}$ given in Eq. (2.7) is associated with a random variable X. The transition matrix $\mathbf{P}_t$ contains elements $\mathbf{P}_t(i,j) = Prob(X(t) = j | X(0) = i)$ which is the probability of being in phase j at time t, given that the initial phase is i. These probabilities are given by $\mathbf{P}_t = e^{\mathbf{Q}t}$, where

$$e^{\mathbf{Q}t} = \begin{bmatrix} e^{\mathbf{D}_0 t} & \mathbf{1} - e^{\mathbf{D}_0 t}\mathbf{1} \\ \mathbf{0} & 1 \end{bmatrix}. \tag{2.13}$$

For the distribution of the time till absorption we have that

$$\begin{aligned} F(t) &= Prob(X(t) = n+1) \\ &= \sum_{1\leq i\leq n+1} Prob(X(0) = i)\, Prob(X(t) = n+1 | X(0) = i) \\ &= \sum_{1\leq i\leq n+1} \boldsymbol{\pi}(i)\mathbf{P}_t(i, n+1) \\ &\stackrel{(2.13)}{=} \boldsymbol{\pi}\mathbf{1} - e^{\mathbf{D}_0 t}\mathbf{1} = 1 - \boldsymbol{\pi} e^{\mathbf{D}_0 t}\mathbf{1}, \end{aligned}$$

holds, which proves Eq. (2.11). Equation 2.12 then follows from $F'(t) = -\boldsymbol{\pi}\frac{d}{dt}\mathbf{P}_t\mathbf{1} = -\boldsymbol{\pi}e^{\mathbf{D}_0 t}\mathbf{D}_0\mathbf{1} = \boldsymbol{\pi}e^{\mathbf{D}_0 t}\mathbf{d}_1$ using $\mathbf{d}_1 = -\mathbf{D}_0\mathbf{1}$.

The expected total time spent in phase j before absorption, given that the initial phase is i equals $-\mathbf{D}_0^{-1}(i,j)$. The ith moment of a PHD is derived from the moment matrix $\mathbf{M} = -\mathbf{D}_0^{-1}$:

$$\mu_i = E[X^i] = i!\boldsymbol{\pi}\mathbf{M}^i\mathbf{1}. \tag{2.14}$$

The event rate is obtained as

$$\lambda = \frac{1}{E[X]} = \frac{1}{\boldsymbol{\pi}\mathbf{M}\mathbf{1}}. \tag{2.15}$$

The squared coefficient of variation equals

$$C^2 = \frac{E[X^2]}{(E[X])^2} - 1 = \frac{2\boldsymbol{\pi}(\mathbf{M})^2\mathbf{1}}{(\boldsymbol{\pi}\mathbf{M}\mathbf{1})^2} - 1. \tag{2.16}$$

Continuous PHDs are dense in the class of distributions on $\mathbb{R}_{\geq 0}$. That is, any distribution with a strictly positive density in $(0,\infty)$ can be approximated arbitrarily close by a PHD (see, e.g. [132]).

2.2 Similarity and Equivalence

It is well known that the matrix representation of a PHD is not unique. Since a PHD of order n is determined by at most $2n-1$ independent parameters [130, 132], but the matrix representation $(\boldsymbol{\pi}, \mathbf{D}_0)$ has n^2+n-1 parameters,[1] this representation is highly redundant, implying that in general, a PH distribution has infinitely many representations with the same order and identical cdf (distribution of the time till absorption) but with different representations of matrix $\mathbf{D}_0$ and vector $\boldsymbol{\pi}$ [129]. Moreover, PHDs $(\boldsymbol{\pi}', \mathbf{D}_0')$ with order $m \neq n$ exist that describe the same distribution as $(\boldsymbol{\pi}, \mathbf{D}_0)$.

Example 2.2. Consider the three PHDs PH_A, PH_B and PH_C defined as

$$\boldsymbol{\pi}^{(A)} = (0.3, 0.3, 0.4) \qquad \mathbf{D}_0^{(A)} = \begin{bmatrix} -4 & 1 & 2 \\ 1 & -4 & 2 \\ 0 & 0 & -1 \end{bmatrix},$$

$$\boldsymbol{\pi}^{(B)} = (0.21, 0.39, 0.4) \qquad \mathbf{D}_0^{(B)} = \begin{bmatrix} -4.3 & 1.3 & 2 \\ 0.7 & -3.7 & 2 \\ 0 & 0 & -1 \end{bmatrix},$$

$$\boldsymbol{\pi}^{(C)} = (0.6, 0.4) \qquad \mathbf{D}_0^{(C)} = \begin{bmatrix} -3 & 2 \\ 0 & -1 \end{bmatrix}.$$

All three representations describe the same PHD. We will continue this example throughout this section and show how one representation can be transformed into another.

Depending on the purpose the PHD is used for, different representations are favorable. If the parameters of a PHD should be estimated (this will be treated in Chap. 3) one is usually interested in canonical forms with only the minimal numbers of parameters required to describe the PHD. Canonical forms are only known for a subclass of PHDs and will be introduced in Sect. 2.3.3. As we will see later (cf. Chap. 5) several approaches for estimating the parameters of a Markovian Arrival Process (MAP) start with a PHD that is expanded into a MAP and the representation of this PHD has a large impact on the flexibility and quality of the estimation approach for the MAP. When the PHD is part of a larger model that should be analyzed numerically, the state space explosion which makes models intractable because of their size, becomes a major problem. In these cases, one is

---

[1] As mentioned in Sect. 2.1.2 we assume that the point mass at zero, i.e., the probability of starting in the absorbing state is 0. If the absorbing state may have an initial probability greater than zero the number of independent parameters increases to $2n$ and the matrix representation has n^2+n parameters.

usually interested in finding a smaller representation of the PHD to diminish the problem. From these examples it is obvious, that there is a need to transform the representation of a PHD into an equivalent representation (either of the same or a smaller order) to be able to find the most adequate representation for a specific application. Several approaches have been proposed for this task. In Sect. 2.2.1 we will give a brief overview on transformations that preserve the order of the PHD, while Sect. 2.2.2 deals with techniques to reduce the number of states. The transformations presented here do not make any assumptions on the structure of $(\boldsymbol{\pi}, \mathbf{D}_0)$ and can be applied to any PHD. For certain subclasses with specific restrictions on $(\boldsymbol{\pi}, \mathbf{D}_0)$ (in particular acyclic PHDs with an upper triangular matrix $\mathbf{D}_0$) further transformations exist and will be treated in Sect. 2.3.3.

2.2.1 Similarity Transformations

From linear algebra it is well known that two matrices $\mathbf{C}$ and $\mathbf{D}$ are similar if there exists a non-singular matrix $\mathbf{B}$ such that $\mathbf{C} = \mathbf{B}\mathbf{D}\mathbf{B}^{-1}$ (or equivalently $\mathbf{C}\mathbf{B} = \mathbf{B}\mathbf{D}$) [119]. Similar matrices share many properties like, e.g. eigenvalues or the characteristic polynomial. This concept of similarity can be used to define equivalence for PH distributions [129, 155].

Let $(\boldsymbol{\pi}, \mathbf{D}_0)$ and $(\boldsymbol{\pi}', \mathbf{D}'_0)$ be two PHDs of the same order with cumulative distribution functions $F(x) = 1 - \boldsymbol{\pi} e^{\mathbf{D}_0 x}\mathbf{1}$ and $G(x) = 1 - \boldsymbol{\pi}' e^{\mathbf{D}'_0 x}\mathbf{1}$, respectively. Let $\mathbf{B}$ be a non-singular matrix with the additional requirement $\mathbf{B}^{-1}\mathbf{1} = \mathbf{1}$. It was shown in [155] that the two PHDs are equivalent if $\boldsymbol{\pi}' = \boldsymbol{\pi}\mathbf{B}$ and $\mathbf{D}'_0 = \mathbf{B}^{-1}\mathbf{D}_0\mathbf{B}$ holds for a matrix $\mathbf{B}$ that fulfills the requirements above. The equivalence follows immediately from

$$G(x) = 1 - \boldsymbol{\pi}' e^{\mathbf{D}'_0 x}\mathbf{1} = 1 - \boldsymbol{\pi}\mathbf{B} e^{\mathbf{B}^{-1}\mathbf{D}_0\mathbf{B}x}\mathbf{B}^{-1}\mathbf{1} = 1 - \boldsymbol{\pi}\mathbf{B}\mathbf{B}^{-1} e^{\mathbf{D}_0 x}\mathbf{B}\mathbf{B}^{-1}\mathbf{1} = 1 - \boldsymbol{\pi} e^{\mathbf{D}_0 x}\mathbf{1} = F(x).$$

It is obvious that for a given PHD $(\boldsymbol{\pi}, \mathbf{D}_0)$ the similarity transformation described above does not result in a valid PHD $(\boldsymbol{\pi}', \mathbf{D}'_0)$ for all matrices $\mathbf{B}$. However, if two PHDs are given, it is easy to verify if the distributions are equivalent by finding a matrix $\mathbf{B}$ that transforms one representation into the other.

Example 2.3. We will use the similarity transformation with a matrix $\mathbf{B}$ to show that PH_A and PH_B describe the same distribution, i.e. $\boldsymbol{\pi}^{(B)} = \boldsymbol{\pi}^{(A)}\mathbf{B}$ and $\mathbf{D}_0^{(B)} = \mathbf{B}^{-1}\mathbf{D}_0^{(A)}\mathbf{B}$ has to hold. We can solve the equations and obtain

$$\mathbf{B} = \begin{bmatrix} 0.7 & 0.3 & 0 \\ 0 & 1 & 0 \\ 0 & 0 & 1 \end{bmatrix}$$

which proves that PH_A and PH_B are indeed equivalent.

2.2.2 Lumping and General Equivalence

Lumping is a technique to reduce the size of state space of a Markov chain [29]. The basic idea is to divide the state space into partitions and to represent each partition by a single state. More formally, we divide the state space $\mathcal{S} = \{1, \cdots, n\}$ into partitions $\Omega = \{\Omega_1, \cdots, \Omega_N\}$ such that

$$\Omega_I \subseteq \mathcal{S}, \qquad \Omega_I \neq \emptyset, \qquad \Omega_I \cap \Omega_J = \emptyset, \qquad \cup_{I=1}^{N} \Omega_I = \mathcal{S}.$$

If we assume that the states belonging to each Ω_I are grouped together we may write the generator matrix $\mathbf{Q}$ of the CTMC as

$$\mathbf{Q} = \begin{bmatrix} \mathbf{Q}_{1,1} & \cdots & \mathbf{Q}_{1,N} \\ \vdots & \ddots & \vdots \\ \mathbf{Q}_{N,1} & \cdots & \mathbf{Q}_{N,N} \end{bmatrix}$$

where submatrix $\mathbf{Q}_{I,J}$ contains the transitions from Ω_I to Ω_J. Additionally, $\boldsymbol{\pi}_I$ is a subvector of $\boldsymbol{\pi}$ that contains the initial probabilities of all states belonging to Ω_I. The partition Ω can be used to construct an aggregated Markov chain by substituting each partition group Ω_I by a single state. Let $q_{i,\Omega_J} = \sum_{j \in \Omega_J} \mathbf{Q}(i,j)$ be the sum of the transition rates from state i into any of the states from Ω_J. If for every pair of partition groups Ω_I and Ω_J, q_{i,Ω_J} is equal for all $i \in \Omega_I$ the Markov chain is lumpable [93] and we may construct the aggregated Markov chain as follows. The matrix $\hat{\mathbf{Q}}$ of the aggregated chain is constructed using a collector matrix $\mathbf{V}$ and a distributor matrix $\mathbf{W}$, i.e. $\hat{\mathbf{Q}} = \mathbf{WQV}$, where $\mathbf{V}$ is a $n \times N$ matrix with $\mathbf{V}(i,I) = 1$ if $i \in \Omega_I$ and 0 otherwise and $\mathbf{W}$ is a non-negative $N \times n$ matrix with unit row sums that contains in row i equal probabilities for states in Ω_I and 0 elsewhere. It is easy to show that $\mathbf{WV} = \mathbf{I}$ holds.

Example 2.4. Consider PHD PH_A with $\boldsymbol{\pi}^{(A)} = (0.3, 0.3, 0.4)$ and

$$\mathbf{D}_0^{(A)} = \begin{bmatrix} -4 & 1 & 2 \\ 1 & -4 & 2 \\ 0 & 0 & -1 \end{bmatrix}.$$

The complete generator matrix $\mathbf{Q}$ is given by

$$\mathbf{Q} = \begin{bmatrix} -4 & 1 & 2 & 1 \\ 1 & -4 & 2 & 1 \\ 0 & 0 & -1 & 1 \\ 0 & 0 & 0 & 0 \end{bmatrix}.$$

We partition the states into three groups, i.e. $\Omega_1 = \{1,2\}, \Omega_2 = \{3\}$ and $\Omega_3 = \{4\}$ contains the absorbing state. It is easy to verify that the q_{i,Ω_J} are equal for the partition groups. The matrices $\mathbf{V}$ and $\mathbf{W}$ are then given by

$$\mathbf{V} = \begin{bmatrix} 1 & 0 & 0 \\ 1 & 0 & 0 \\ 0 & 1 & 0 \\ 0 & 0 & 1 \end{bmatrix} \quad \text{and} \quad \mathbf{W} = \begin{bmatrix} 1/2 & 1/2 & 0 & 0 \\ 0 & 0 & 1 & 0 \\ 0 & 0 & 0 & 1 \end{bmatrix}.$$

The generator matrix of the aggregated chain is obtained from

$$\hat{\mathbf{Q}} = \mathbf{WQV} = \begin{bmatrix} -3 & 2 & 1 \\ 0 & -1 & 1 \\ 0 & 0 & 0 \end{bmatrix},$$

which is the generator matrix of PH_C. The initial probability vector of the aggregate can be obtained by adding all initial probabilities of the states in each Ω_I, which is the same as computing $\hat{\boldsymbol{\pi}} = \boldsymbol{\pi}\mathbf{V}$.

In most cases PHDs are not lumpable at all or only very few states can be saved by this aggregation approach.

Note, that from an algebraic point of view the similarity transformation from the previous section and lumping are defined in a similar way, i.e. in both cases the (sub-)generator matrix of the PHD is multiplied by two matrices with row sums equal to one.

Consequently, the most general definition of equivalence between PHDs that has been proposed in [36–38] generalizes the similarity transformation from [155] and lumpability and introduces a description of equivalence between PHDs of the same and different orders. Let $(\boldsymbol{\pi}, \mathbf{D}_0)$ and $(\boldsymbol{\pi}', \mathbf{D}_0')$ be two PHDs of order m and n $(m \geq n)$, respectively. Then the two representations are equivalent if there exists a $m \times n$ matrix $\mathbf{V}$ such that $\mathbf{V1} = \mathbf{1}$, $\mathbf{D}_0\mathbf{V} = \mathbf{V}\mathbf{D}_0'$ and $\boldsymbol{\pi}\mathbf{V} = \boldsymbol{\pi}'$. The same holds for a $n \times m$ matrix $\mathbf{W}$ with $\mathbf{W1} = \mathbf{1}$, $\mathbf{WD}_0 = \mathbf{D}_0'\mathbf{W}$ and $\boldsymbol{\pi} = \mathbf{W}\boldsymbol{\pi}'$. This definition is not only valid for PHDs but also holds for ME distributions [111], which we will not cover throughout this work.

2.3 Acyclic Phase-Type Distributions

We introduce some basic properties of acyclic PHDs.

Definition 2.5. If the transition rate matrix $\mathbf{D}_0$ can be transformed into an upper (or lower) triangular matrix by symmetric permutations of rows and columns the PHD is called an acyclic phase-type distribution (APHD).

Since the matrix $\mathbf{D}_0$ is of an upper triangular form, phase i can only be connected with phase j if $j > i$. Consequently, each transient phase is visited at most once before absorption. The matrix representation $(\boldsymbol{\pi}, \mathbf{D}_0)$ has $(n^2+n)/2$ parameters for the upper triangular matrix and $n-1$ free parameters for the initial distribution vector. APHDs are the largest subclass of PHDs for which canonical representations exist (cf. Sect. 2.3.3).

APHDs can be divided into various subclasses depending on the structure of $\mathbf{D}_0$ and $\boldsymbol{\pi}$. In the following we will give an overview of these distributions.

2.3.1 *Erlang and Hyper-Erlang Distributions*

Since PHDs can be considered as a natural generalization of the exponential and Erlang distributions we introduce them in this section. The exponential distribution is completely characterized by its rate parameter λ and is the simplest case of a PHD with a single transient state as shown in Fig. 2.4a. By the fact that $\boldsymbol{\pi} = [1]$, the only transient phase is visited before absorption. The corresponding infinitesimal generator matrix includes $\mathbf{D}_0 = [-\lambda]$ and the exit vector $\mathbf{d}_1 = [\lambda]$ as shown in Eq. (2.17) in Fig. 2.4b.

a

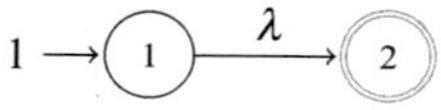

b

$$\mathbf{Q} = \begin{bmatrix} -\lambda & \lambda \\ 0 & 0 \end{bmatrix} \tag{2.17}$$

Fig. 2.4 Markovian representation of the exponential distribution. (**a**) An exponential distribution with parameter λ, and 2 being an absorbing state. (**b**) The infinitesimal generator $\mathbf{Q}$

The exponential distribution has density

$$f(x) = \lambda e^{-\lambda x} \text{ for } x \geq 0, \tag{2.18}$$

and its distribution function is given by

$$F(x) = 1 - e^{-\lambda x} \text{ for } x \geq 0. \tag{2.19}$$

Foremost it is the only continuous distribution with the memoryless property

$$Prob(X > t+s | X > t) = Prob(X > s) \text{ for all } t, s \geq 0.$$

The mean is given by $E[X] = \frac{1}{\lambda}$ and the variance is $VAR[X] = \frac{1}{\lambda^2}$.

Erlang introduced in [53] the representation of distributions as a sum of n exponential phases with the same intensity λ. Consider n mutually independent, exponentially distributed random variables X_i with parameter $\lambda > 0$, $1 \leq i \leq n$. If we define a random variable Y as $Y = \sum_{1 \leq i \leq n} X_i$, then it has an Erlang distribution denoted by $E(n, \lambda)$, and its density is given by

$$f(x) = \frac{\lambda^n}{(n-1)!} x^{n-1} e^{-\lambda x} \text{ for } x \geq 0. \tag{2.20}$$

The distribution function is defined by

$$F(x) = 1 - \sum_{i=0}^{n-1} \frac{(\lambda x)^i}{i!} e^{-\lambda x} \text{ for } x \geq 0. \tag{2.21}$$

The ith moment of the Erlang distributed random variable Y is given by

$$E[Y^i] = \frac{(n+i-1)!}{(n-1)!} \frac{1}{\lambda^i}. \tag{2.22}$$

Thus, the mean of Y is $E[Y] = \frac{n}{\lambda}$ and the variance equals $VAR[Y] = \frac{n}{\lambda^2}$.

The underlying Markov process can be described by the infinitesimal generator matrix given in Eq. (2.23), where the initial probability vector is $\boldsymbol{\pi} = [1, 0, \ldots, 0]$ as shown in Fig. 2.5a.

a

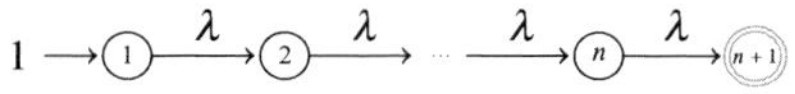

b

$$\mathbf{D}_0 = \begin{bmatrix} -\lambda & \lambda & \ldots & 0 & 0 \\ 0 & -\lambda & \ldots & 0 & 0 \\ \ldots & \ldots & \ddots & \ldots & \ldots \\ 0 & 0 & \ldots & -\lambda & \lambda \\ 0 & 0 & \ldots & 0 & -\lambda \end{bmatrix} \tag{2.23}$$

Fig. 2.5 Erlang representation of a PHD. (**a**) A graphical representation of the Erlang(n,λ) PHD. (**b**) The sub generator $\mathbf{D}_0$

The Markov process must start in phase 1 and traverses through the successive states until it reaches the absorbing state $n+1$. Thus, the time to absorption described by Y is the summation of all holding times which are identically exponentially distributed with parameter λ. The Erlang distribution $E(n,\lambda)$ has a squared coefficient of variation of $C^2 = n^{-1}$ which is less than one for $n > 1$. Distributions with a coefficient of variation greater than one can be modeled as finite mixtures of exponential distributions.

Remark 2.1. The Erlang distribution with mean m has variance $\frac{m^2}{n}$ and thus tends to be deterministic for $n \to \infty$. In this case the squared coefficient of variation is close to zero. This coefficient is used to express the variance of the random variable relative to its average value. Consequently, the Erlang distribution can be used to approximate deterministic distributions.

As a next example we consider the hypo-exponential distribution which is a generalized Erlang distribution. Consider a set of exponential distributions $F_i(\cdot)$ with

$$F_i(x) = 1 - e^{-\lambda(i)x} \text{ for } x \geq 0,\ 1 \leq i \leq n,$$

and where rates $\lambda(1),\ldots,\lambda(n)$ are not necessarily identical. Consequently we have

$$f_i(x) = \lambda(i)e^{-\lambda(i)x} \text{ for } x \geq 0.$$

The hypo-exponential distribution is then characterized by the number n and the set of parameters $\lambda(i)$. Its density function is given by

$$f(x) = \sum_{i=1}^{n} \left(\prod_{j=1, j\neq i}^{n} \frac{\lambda(j)}{\lambda(j)-\lambda(i)} \right) f_i(x) \text{ for } x \geq 0,\ \lambda(i) \neq \lambda(j) \text{ for } i \neq j. \tag{2.24}$$

We obtain the mean of the hypo-exponentially distributed random variable as $E[X] = \sum_{i=1}^{n} \frac{1}{\lambda(i)}$ and the variance as $VAR[X] = \sum_{i=1}^{n} \frac{1}{\lambda(i)^2}$. If all parameters $\lambda(i)$ are equal, we obtain the Erlang distribution since it is the convolution of n identical exponentials. The sub generator of the hypo-exponential distribution is shown in Eq. (2.25) and Fig. 2.6b. The initial distribution vector is $\boldsymbol{\pi} = [1, 0, \ldots, 0]$ as shown in Fig. 2.6a.

a

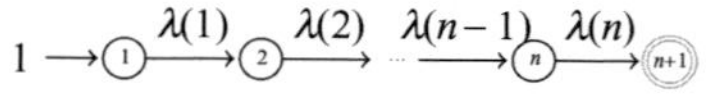

b

$$\mathbf{D}_0 = \begin{bmatrix} -\lambda(1) & \lambda(1) & \ldots & 0 & 0 \\ 0 & -\lambda(2) & \ldots & 0 & 0 \\ \ldots & \ldots & \ddots & \ldots & \ldots \\ 0 & 0 & \ldots & -\lambda(n-1) & \lambda(n-1) \\ 0 & 0 & \ldots & 0 & -\lambda(n) \end{bmatrix} \tag{2.25}$$

Fig. 2.6 The hypo-exponential distribution. (**a**) Markovian representation of the hypo-exponential distribution. In particular $\lambda(1),\ldots,\lambda(n)$ are not necessarily identical. (**b**) The sub generator $\mathbf{D}_0$

The hyper-exponential distribution is a convex mixture of n exponential distributions and is visualized in Fig. 2.7. The density is given by

$$f(x) = \sum_{i=1}^{n} \pi(i)\lambda(i)e^{-\lambda(i)x} \text{ for } x \geq 0, \tag{2.26}$$

where $\pi(i) > 0$ for all phases i and $\sum_{i=1}^{n} \pi(i) = 1$. The distribution function of a hyper-exponentially distributed random variable X is defined as

$$F(x) = \sum_{i=1}^{n} \pi(i)(1 - e^{-\lambda(i)x}) \text{ for } x \geq 0. \tag{2.27}$$

The first moment is obtained as $E[X] = \sum_{i=1}^{n} \frac{\pi(i)}{\lambda(i)}$ and its variance is given by

$$VAR[X] = 2\sum_{i=1}^{n} \frac{\pi(i)}{\lambda(i)^2} - \left(\sum_{i=1}^{n} \frac{\pi(i)}{\lambda(i)} \right)^2. \tag{2.28}$$

The intensity matrix of the hyper-exponential distribution has a form shown in Eq. (2.29) shown in Fig. 2.7b, and the initial probability vector is $\boldsymbol{\pi} = [\pi(1), \pi(2), \ldots, \pi(n)]$. A graphical representation of the corresponding Markovian process is

a

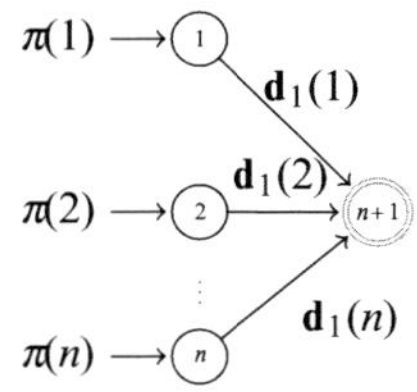

b

$$\mathbf{D}_0 = \begin{bmatrix} -\lambda(1) & 0 & \ldots & 0 & 0 \\ 0 & -\lambda(2) & \ldots & 0 & 0 \\ \ldots & \ldots & \ddots & \ldots & \ldots \\ 0 & 0 & \ldots & -\lambda(n-1) & 0 \\ 0 & 0 & \ldots & 0 & -\lambda(n) \end{bmatrix} \tag{2.29}$$

Fig. 2.7 The hyper-exponential distribution. (**a**) A graphical representation of the hyper-exponential distribution. (**b**) The sub generator $\mathbf{D}_0$

visualized in Fig. 2.7a. By the fact, that the Markov process can start in each phase an additional dispersion is introduced which leads to the squared coefficient of variation greater than one; and equal to one for the exponential case $n = 1$.

$$C^2 = \frac{E[Y^2]}{(E[Y])^2} - 1 = 2\frac{\sum_{i=1}^{n} \frac{\pi(i)}{\lambda(i)^2}}{\left(\sum_{i=1}^{n} \frac{\pi(i)}{\lambda(i)}\right)^2} - 1 \tag{2.30}$$

A hyper-Erlang distribution denoted as HErD [57], is a mixture of m mutually independent Erlang distributions weighted with the initial probabilities $\pi(1), \ldots, \pi(m)$, where $\pi(i) \geq 0$, and the vector $\boldsymbol{\pi}$ is stochastic, i.e. $\sum_{i=1}^{m} \pi(i) = 1$. Let s_i denote the number of phases of the ith Erlang distribution. Then the density is

$$f(x) = \sum_{i=1}^{m} \pi(i) \frac{(\lambda(i)x)^{s_i - 1}}{(s_i - 1)!} \lambda(i) e^{-\lambda(i)x} \text{ for } x \geq 0, \tag{2.31}$$

and its distribution function is given by

$$F(x) = 1 - \sum_{i=1}^{m} \pi(i) \sum_{j=0}^{s_i - 1} \frac{(\lambda(i)x)^j}{j!} e^{-\lambda(i)x} \text{ for } x \geq 0. \tag{2.32}$$

The ith moment can be obtained as

$$E[Y^i] = \sum_{j=1}^{m} \pi(i) \frac{(s_j + i - 1)!}{(s_j - 1)!} \frac{1}{\lambda(j)^i}. \tag{2.33}$$

The state space of HErD consists of $\sum_{i=1}^{m} s_i$ transient and one absorbing state. For $m = 1$ we have a single Erlang distribution $E(s_1, \lambda(1))$ and the case that

$s_i = 1$ for all $1 \le i \le m$ represents a hyper-exponential distribution. The underlying Markov chain of the HErD can be described by the infinitesimal generator given in Eq. (2.34) as shown in Fig. 2.8b, which has the matrices $\mathbf{Q}_i$ on its diagonal. Matrix $\mathbf{Q}_i$ represents the infinitesimal generator of the ith Erlang branch and its form is given in Eq. (2.23). The initial distribution vector is $\boldsymbol{\pi} = [\pi(1), 0, \ldots, 0, \pi(2), 0, \ldots, \pi(m), 0, \ldots, 0]$.

a

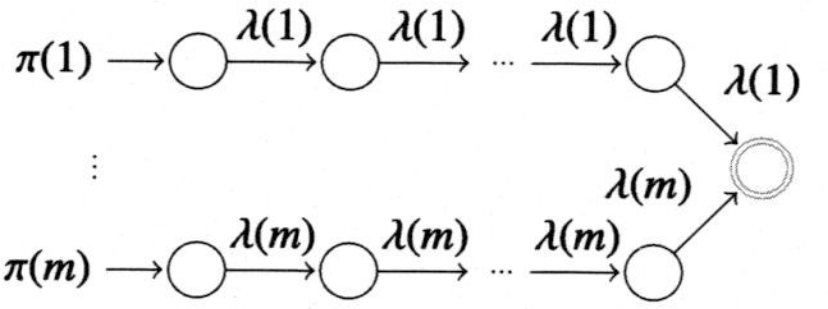

b

$$\mathbf{Q} = \begin{bmatrix} \mathbf{Q}_1 & 0 & \ldots & 0 \\ 0 & \mathbf{Q}_2 & \ldots & 0 \\ \ldots & \ldots & \ddots & \ldots \\ 0 & 0 & \ldots & \mathbf{Q}_m \end{bmatrix} \tag{2.34}$$

Fig. 2.8 Symbolic representation of the HErD. (**a**) A graphical representation of the HErD. (**b**) The infinitesimal generator **Q**

Recall that an Erlang distribution with s phases is defined as the sum of s independent identically distributed random variables. Therefore, a HErD is constructed from a mixture of sums of exponential distributions.

2.3.2 *Coxian Distributions*

Coxian distributions can be considered as a mixture of hypo- and hyper-exponential distributions [45, 157]. The initial distribution vector is given by $\boldsymbol{\pi} = [1, 0, \ldots, 0]$. After starting in phase 1 the process traverses through the n successive phases with possibly different rates $\lambda(i)$. From the phase i the transition to the next phase $i+1$ can occur with probability g_i or the absorbing state is reached with the complementary probability $1 - g_i$.

a

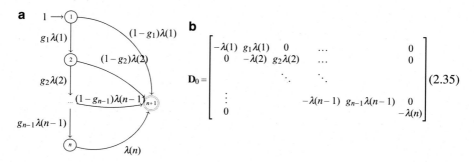

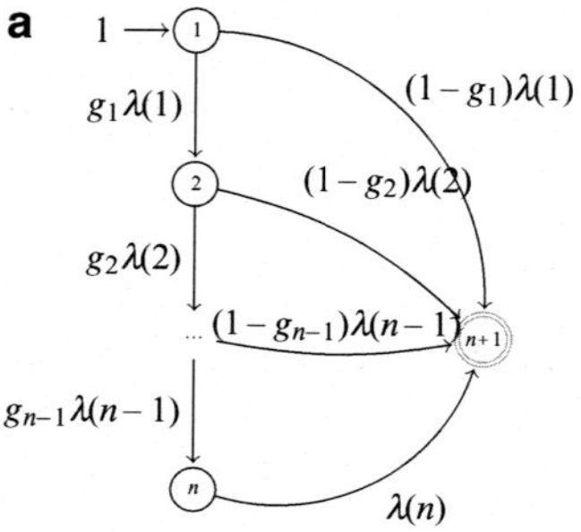

b

$$\mathbf{D}_0 = \begin{bmatrix} -\lambda(1) & g_1\lambda(1) & 0 & \ldots & & 0 \\ 0 & -\lambda(2) & g_2\lambda(2) & \ldots & & 0 \\ & & \ddots & \ddots & & \\ \vdots & & & -\lambda(n-1) & g_{n-1}\lambda(n-1) & 0 \\ 0 & & & & & -\lambda(n) \end{bmatrix} \tag{2.35}$$

Fig. 2.9 Symbolical representation of the Coxian distribution. (**a**) A graphical representation of the Coxian distribution. (**b**) The sub generator $\mathbf{D}_0$

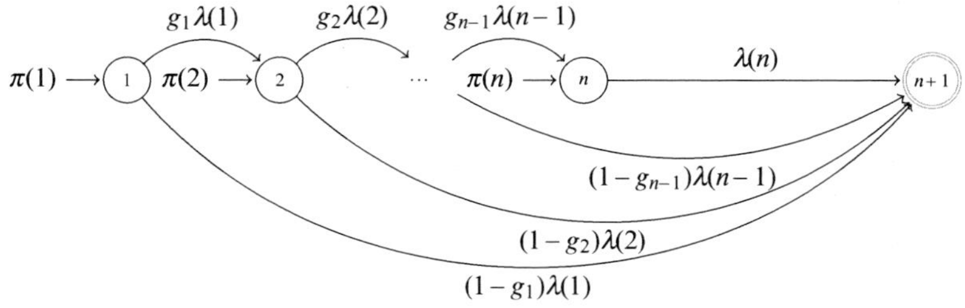

Fig. 2.10 The generalized Coxian PHD

The matrix representation of the Coxian distribution is given by Eq. (2.35) presented in Fig. 2.9b. A generalized Coxian distribution extends Coxian distribution with a random initial vector $\boldsymbol{\pi}$, so that each state can be an entry state. Thus exhibiting low or high variability. A CTMC representation of the Coxian PHD is visualized in Fig. 2.9a, and of the generalized Coxian PHD in Fig. 2.10.

2.3.3 *Canonical Representations*

We now consider distinct canonical representations of an APHD, i.e. a triangular matrix $\mathbf{D}_0$ with nonzero diagonal elements. The analysis of APHDs in [47] has revealed that the cdf of n-phase APHDs has at most $2n-1$ degrees of freedom. On the other hand, the number of parameters needed to specify an APHD with an upper triangular matrix representation is $(n^2+n)/2$ for the matrix $\mathbf{D}_0$ and $n-1$ parameters for the initial distribution vector. Since the representation of the cdf by a tuple $(\boldsymbol{\pi}, \mathbf{D}_0)$ is highly redundant, minimal and cdf retentive representations have been developed. We will now pose the canonical representations with $2n-1$ free parameters regarding the APHD (see Sect. 2.3). The key idea of the approach from [47] is to express an APHD in terms of its elementary series, i.e. paths from an initial to the absorbing state.

Definition 2.6. For an APHD of order n an elementary series of order $m \leq n$ is defined as a series of the form

$$ES = <\lambda(i_1)\lambda(i_2)\ldots\lambda(i_m)>,$$

where $i_1, i_2, \ldots, i_{m-1}, i_m$ is a sequence of states along the acyclic path from an initial state to the absorbing state such that

$$\mathbf{D}_0(i_k, i_{k+1}) \neq 0,\ k = 1, 2, \ldots, m.$$

and $\mathbf{d}_1(i_m) > 0$, i.e. a transition from the last state of the series i_m to the absorbing state is possible.

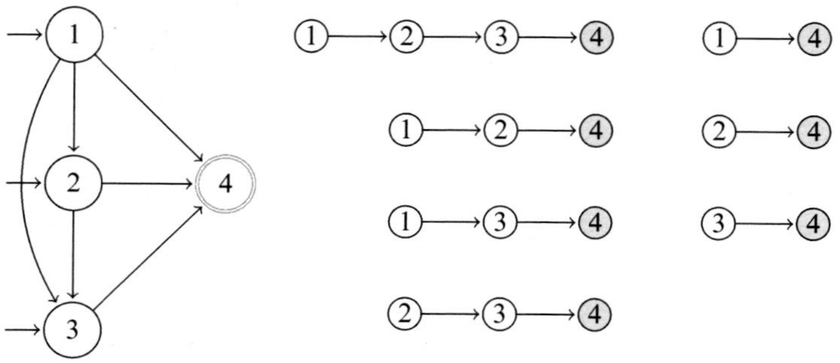

Fig. 2.11 An acyclic 3-phase PHD and its elementary series

The rate between two states of an elementary series is given by the total event rate of the first of the two states, i.e. by the corresponding diagonal entry of $-\mathbf{D}_0$. One can also introduce a path as a sequence of connected states by inspection of the transition diagram corresponding to $(\boldsymbol{\pi}, \mathbf{D}_0)$. The number of possible elementary series in a n-phase APHD is bounded by $2^n - 1$.

Figure 2.11 shows a 3-phase APHD with its elementary series.

Now consider the following identity: Given two positive real numbers λ and μ, with $\lambda \le \mu$, one obtains

$$\frac{\lambda}{s+\lambda} = \tau \frac{\mu}{s+\mu} + (1-\tau)\frac{\lambda\mu}{(s+\lambda)(s+\mu)}, \tag{2.36}$$

where $\tau = \frac{\lambda}{\mu} \in (0,1]$ represents the probability for the path till absorption containing only a phase with the transition rate μ. Consequently, $(1-\tau)$ is the remaining probability for the elementary series containing rates λ and μ, i.e. there are successive phases with transition rates λ and μ. Whenever it holds that $\lambda \le \mu$, we obtain an ascending ordering of the transition rates. Accordingly, we indicate that the cdf of an elementary series has the Laplace transform

$$F(s) = \frac{\lambda(i_1)\lambda(i_2)\ldots\lambda(i_{m-1})}{s(s+\lambda(i_1))(s+\lambda(i_2))\ldots(s+\lambda(i_{m-1}))}, \tag{2.37}$$

e.g. the cdf of the elementary series consisting only of the rate λ has the Laplace transform $F(s) = \frac{\lambda}{s+\lambda}$. With these observations in mind, we can now retain that an elementary series for some phase with transition rate λ can be substituted by a mixture of two elementary series, one containing a phase with transition rate $\mu > \lambda$, and the other containing both phases with the rates λ and μ. This substitution step is visualized in Fig. 2.12 and by repeated use of Eq. (2.36) the elementary series can be transformed to a mixture of basic series, which can be specified now.

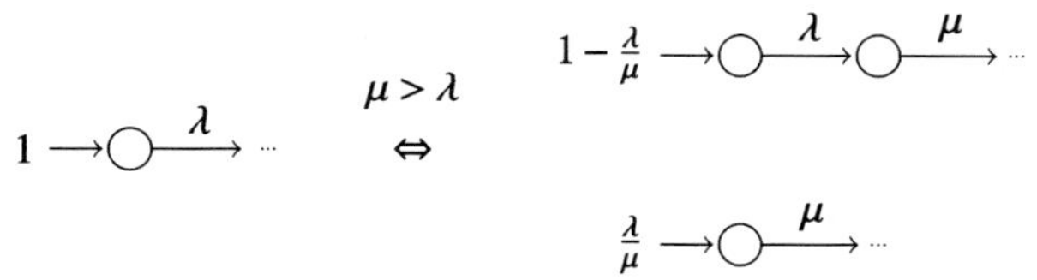

Fig. 2.12 Substitution step for the exponential distribution

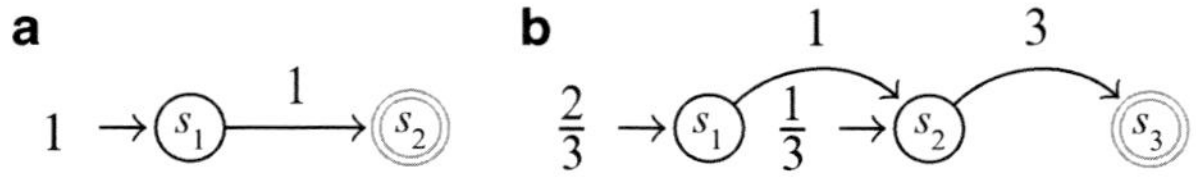

Fig. 2.13 Equivalent representation of the exponential distribution (**a**) Series for an exponential distribution. (**b**) Series for a 2 phase APHD

Example 2.5. The 3-phase PHD shown in Fig. 2.13b has the Laplace transform $F(s) = \frac{2}{3}\frac{1\cdot 3}{(s+1)(s+3)} + \frac{1}{3}\frac{3}{(s+3)} = \frac{2}{(s+1)(s+3)} + \frac{1}{s+3} = \frac{1}{s+1}$. Thus, this PHD is cdf-equivalent to the former PHD shown in Fig. 2.13a since it contains the elementary series $<\mu>$ and $<\lambda\mu>$ with $\lambda = 1$ and $\mu = 3$.

Definition 2.7. For n positive real numbers $0 < \lambda(1) \leq \lambda(2) \leq \ldots \leq \lambda(n)$ the basic series (BS) are defined as the tuples of $i,\ldots,n$ transient states that determine the acyclic path till absorption [47]. Each basic series can be represented by the notation,

$$BS_i = <\lambda(i)\ldots\lambda(n-1)\lambda(n)>.$$

The Markov process can start in a phase with transition rate $\lambda(i)$ and follows the path till absorption. Thus, the exit rate of each path is $\lambda(n)$ and according to the substitution step the canonical ordering of $\lambda(i)$'s is incorporated.

It is now easy to see, that each basic series BS_i describes a hypo-exponential distribution (cf. Eq. (2.25) visualized in Fig. 2.6b) with the corresponding distribution of the time till absorption, i.e., it is the convolution of $(n-i+1)$ exponentials with parameters $\lambda(i),\ldots,\lambda(n-1),\lambda(n)$. The results show that for an APH, the cdf of each of its elementary series is a mixture of the cdfs of its basic series. Furthermore, the mixture of the cdfs of all elementary series determines the distribution of the time till absorption for the underlying APH, where each elementary series is weighted proportional to its probability. The probability of each elementary series is computed as the product of the transition rates along the corresponding path and the initial probability of the first state of the path. Let $i_1, i_2, \ldots, i_m$ be m states from the jth elementary series of a PHD. Then, the probability of the ES_j is given by

$$\tau_j = \pi(i_1)\frac{\mathbf{D}_0(i_1,i_2)}{-\mathbf{D}_0(i_1,i_1)}\frac{\mathbf{D}_0(i_2,i_3)}{-\mathbf{D}_0(i_2,i_2)}\cdots\frac{\mathbf{D}_0(i_{m-1},i_m)}{-\mathbf{D}_0(i_{m-1},i_{m-1})}\frac{\mathbf{d}_1(i_m)}{-\mathbf{D}_0(i_m,i_m)}, \tag{2.38}$$

where the term $\frac{\mathbf{D}_0(i_k,i_{k+1})}{-\mathbf{D}_0(i_k,i_k)}$, for any $k = 1,\ldots,m-1$, represents the transition probability from the state i_k to the state i_{k+1}. Basic series together with appropriate initial probabilities yield the series canonical form.

Definition 2.8. The series canonical form is defined as a mixture of basic series of an APHD with transition rates in ascending order, i.e. $\lambda(i) \leq \lambda(i+1) \leq \ldots \leq \lambda(n)$. Transitions are only possible from phase i to the neighbor phase $i+1$. There is only one exit state, but all states may be entry states satisfying $\boldsymbol{\pi}(i) \geq 0$, for all $i = 1,\ldots,n$. The matrix representation of the series canonical form is given by Eq. (2.39) in Fig. 2.14a.

a

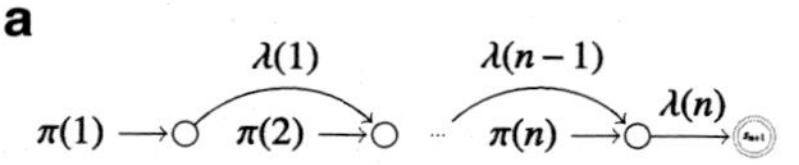

b

$$\mathbf{D}_0 = \begin{bmatrix} -\lambda(1) & \lambda(1) & 0 & \cdots & 0 \\ 0 & -\lambda(2) & \lambda(2) & & \\ & & \ddots & \ddots & \vdots \\ \vdots & & & -\lambda(n-1) & \lambda(n-1) \\ 0 & & & & -\lambda(n) \end{bmatrix} \quad (2.39)$$

Fig. 2.14 PHD in series canonical form. (**a**) The series canonical form, in particular $0 < \lambda(1) \leq \lambda(2) \leq \ldots \leq \lambda(n-1) \leq \lambda(n)$. (**b**) The sub generator $\mathbf{D}_0$

The series canonical form is visualized in Fig. 2.14a and has $2n-1$ degrees of freedom, namely n transition rates and $n-1$ independent initial probabilities [47].

For general PHDs canonical forms are only known for the cases $n = 2$ or $n = 3$. For the case $n = 2$ it has been shown that APHDs and general PHDs are equivalent in the sense that every distribution that is represented as a PHD also has a representation as an APHD [45]. For $n = 3$ a canonical form for general PHDs has been developed in [78] based on earlier results in [66]. The canonical representation is generated by symbolically performing similarity transformations on matrix $\mathbf{D}_0$ and vector $\boldsymbol{\pi}$ using the equivalence of PHDs defined in Sect. 2.2. However, it can be shown that this approach cannot be applied for $n > 3$.

2.4 Properties

Since the class of PHDs is closed under certain operations such as convolutions and finite mixtures, we summarize below some of the basic properties without proof. The presentation is based on results discussed in [103, 127]. Further closure properties of PHDs can be found in [116].

First we consider the sum of two independent random variables of the phase-type. Let $PH_A = (\boldsymbol{\pi}^{(A)}, \mathbf{D}_0^{(A)})$ be of order n, and $PH_B = (\boldsymbol{\pi}^{(B)}, \mathbf{D}_0^{(B)})$ be of order m. Furthermore PH_A is the distribution of the random variable $X^{(A)}$, and PH_B is the distribution of the random variable $X^{(B)}$. Then the sum $X^{(C)} = X^{(A)} + X^{(B)}$ is phase-type distributed. The underlying Markov process can be described by the sub generator matrix given in Eq. (2.40) where the initial probability vector is $\pi^{(C)} = [\pi^{(A)}, \pi^{(A)}(n+1)\pi^{(B)}]$.[2]

$$\mathbf{D}_0^{(C)} = \begin{bmatrix} \mathbf{D}_0^{(A)} & \mathbf{d}_1^{(A)}\pi^{(B)} \\ \mathbf{0} & \mathbf{D}_0^{(B)} \end{bmatrix}. \tag{2.40}$$

In the underlying Markov chain $\{X^{(C)}(t)\}_{t\geq 0}^{\infty}$ the paths of the Markov chains associated with PH_A and PH_B are concatenated such that after traversing the paths of the chain $\{X^{(A)}(t)\}_{t\geq 0}^{\infty}$ the process moves along the paths of the chain $\{X^{(B)}(t)\}_{t\geq 0}^{\infty}$. The resulting number of transient phases in $\{X^{(C)}(t)\}_{t\geq 0}^{\infty}$ is $n+m$. The initial phase is selected according to $\pi^{(A)}$. Then the process moves along the paths of the $\{X^{(A)}(t)\}_{t\geq 0}^{\infty}$ until some entry state of the chain $\{X^{(B)}(t)\}_{t\geq 0}^{\infty}$ is reached. By the fact that $\mathbf{d}_1^{(A)}\pi^{(B)}$ is the rate of entering some initial state of the second chain, the initial distribution of that process remains unchanged. Thus, the absorbing time of the constructed Markov process $\{X^{(C)}(t)\}_{t\geq 0}^{\infty}$ is the sum of the absorbing time of the first and the second Markov chains.

As next operation we consider the convex mixture of PHDs. Let again $PH_A = (\pi^{(A)}, \mathbf{D}_0^{(A)})$ be of order n, and $PH_B = (\pi^{(B)}, \mathbf{D}_0^{(B)})$ be of order m. Furthermore let $F^{(A)}(\cdot)$ and $F^{(B)}(\cdot)$ be the corresponding probability distribution functions. Then let the PHD PH_C be a convex mixture of these distribution functions which is defined as $\alpha F^{(A)}(\cdot) + (1-\alpha)F^{(B)}(\cdot)$, for $0 \geq \alpha \leq 1$. The underlying Markov process has $n+m$ transient phases, the initial probability vector equals $\pi^{(C)} = [\alpha\pi^{(A)}, (1-\alpha)\pi^{(B)}]$, and the sub generator matrix is given in Eq. (2.41).

$$\mathbf{D}_0^{(C)} = \begin{bmatrix} \mathbf{D}_0^{(A)} & \mathbf{0} \\ \mathbf{0} & \mathbf{D}_0^{(B)} \end{bmatrix}. \tag{2.41}$$

Since the states of PH_A, PH_B are disjoint in the process associated with PH_C, the set of the passed states corresponds either to the PH_A or to the PH_B. Thus, the time until absorption is distributed either according to PH_A with probability α, or it is distributed like PH_B with complementary probability.

All order statistics of a finite number of independent PHD random variables, e.g. the kth smallest of a set of random variables, minima, or maxima, are PHDs [103].

We focus on the distribution of the smallest and the largest of two independent random variables $X^{(A)}$, $X^{(B)}$ of the phase-type. The corresponding representation of the PHDs is $PH_A = (\pi^{(A)}, \mathbf{D}_0^{(A)})$ of order n, and $PH_B = (\pi^{(B)}, \mathbf{D}_0^{(B)})$ of order m, respectively. Then the random variable $X^{(C)} = min(X^{(A)}, X^{(B)})$ has a PHD with representation $(\pi^{(C)}, \mathbf{D}_0^{(C)})$. The initial distribution is defined as $\pi^{(C)} = [\pi^{(A)} \otimes \pi^{(B)}]$, and the corresponding sub generator matrix is given in Eq. (2.42).

---

[2] If the case that $\pi^{(A)}(n+1) = 0$, i.e., there is no start in the absorbing state, the random variable $X^{(A)}$ is strictly positive. Then the initial probability vector is given by $\pi^{(C)} = [\pi^{(A)}, \mathbf{0}]$ where $\mathbf{0}$ is the row m-vector of 0's.

$$\mathbf{D}_0^{(C)} = \mathbf{D}_0^{(A)} \otimes \mathbf{I}^{(B)} + \mathbf{I}^{(A)} \otimes \mathbf{D}_0^{(B)} = \mathbf{D}_0^{(A)} \oplus \mathbf{D}_0^{(B)}, \tag{2.42}$$

where $\mathbf{I}^{(A)}$, $\mathbf{I}^{(B)}$ are identity matrices of order n and m, respectively. $\otimes$ and $\oplus$ denote the Kronecker product and Kronecker sum [49, 112] which are defined for two square matrices $\mathbf{A}$ and $\mathbf{B}$ of order n^a and n^b as

$$\mathbf{A} \otimes \mathbf{B} = \begin{bmatrix} \mathbf{A}(1,1)\mathbf{B} & \cdots & \mathbf{A}(1,n^a)\mathbf{B} \\ \vdots & \ddots & \vdots \\ \mathbf{A}(n^a,1)\mathbf{B} & \cdots & \mathbf{A}(n^a,n^a)\mathbf{B} \end{bmatrix} \text{ and } \mathbf{A} \oplus \mathbf{B} = \mathbf{A} \otimes \mathbf{I}^{n^a} + \mathbf{I}^{n^b} \otimes \mathbf{B}. \tag{2.43}$$

In the resulting two-dimensional Markov process of order nm the state space is given by pairs of phases $\{(i,j) : i \in \mathcal{S}_T^{(A)}, j \in \mathcal{S}_T^{(B)}\}$ and the single absorbing state $nm+1$. The underlying chain models the concurrent behavior of the two original processes $\{X^{(A)}(t)\}_{t\geq 0}^{\infty}$, $\{X^{(B)}(t)\}_{t\geq 0}^{\infty}$. Strictly, the expanded chain has been obtained through sequencing of concurrent state transitions in the original chains. For example, it is possible that either the first chain or the second chain alone evolves through the state space till absorption. In one of that cases only one component in the tuple (i, j) changes. If it gets absorbed, this means that the minimum of both PHDs is determined.

We conclude with results for the largest of the two independent random variables $X^{(A)}$, $X^{(B)}$ of the phase-type. The random variable $X^{(C)} = max(X^{(A)}, X^{(B)})$ has a PHD with representation $(\boldsymbol{\pi}^{(C)}, \mathbf{D}_0^{(C)})$, where the initial probability vector is given by $\boldsymbol{\pi}^{(C)} = [\boldsymbol{\pi}^{(A)} \otimes \boldsymbol{\pi}^{(B)}, \boldsymbol{\pi}^{(A)}\boldsymbol{\pi}^{(B)}(m+1), \boldsymbol{\pi}^{(A)}(n+1)\boldsymbol{\pi}^{(B)}]$. The sub generator of the associated Markov chain is given in Eq. (2.44)

$$\mathbf{D}_0^{(C)} = \begin{bmatrix} \mathbf{D}_0^{(A)} \oplus \mathbf{D}_0^{(B)} & \mathbf{I}^{(A)} \otimes \mathbf{d}_1^{(B)} & \mathbf{d}_1^{(A)} \otimes \mathbf{I}^{(B)} \\ \mathbf{0} & \mathbf{D}_0^{(A)} & \mathbf{0} \\ \mathbf{0} & \mathbf{0} & \mathbf{D}_0^{(B)} \end{bmatrix}. \tag{2.44}$$

The expanded Markov chain corresponding to the distribution with representation $(\boldsymbol{\pi}^{(C)}, \mathbf{D}_0^{(C)})$ has been formed as follows. Its state space consists of $nm+n+m$ transient states. The submatrix $\mathbf{D}_0^{(A)} \otimes \mathbf{I}^{(B)} + \mathbf{I}^{(A)} \otimes \mathbf{D}_0^{(B)}$ describes the part where the original processes evolve simultaneously until one of them gets absorbed. This means that the minimum of both PHDs is known and we shall retain the absorbed process. Thus additional states $(n+1,\cdot)$ and $(\cdot, m+1)$ corresponds to the absorption of one of the involved processes. Reaching one of the states $(n+1,\cdot)$ or $(\cdot, m+1)$ the chain evolves according to the remaining process which has been not absorbed yet. The state space contains pairs of phases $\{(i,j) \mid i \in \mathcal{S}_T^{(A)},\ j \in \mathcal{S}_T^{(B)}\} \cup \{(n+1,j) \mid n+1 \in \mathcal{S}^{(A)} \setminus \mathcal{S}_T^{(A)},\ j \in \mathcal{S}_T^{(B)}\} \cup \{(i,m+1) \mid, i \in \mathcal{S}_T^{(A)},\ m+1 \in \mathcal{S}^{(B)} \setminus \mathcal{S}_T^{(B)}\}$, and the absorbing state. Hence all combinations of transitions until absorption of both Markov chains are considered.

a

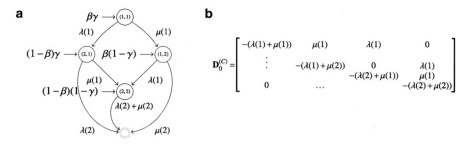

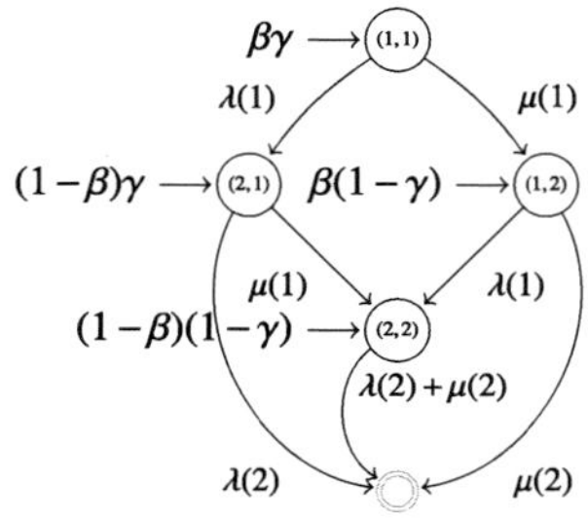

b

$$\mathbf{D}_0^{(C)} = \begin{bmatrix} -(\lambda(1)+\mu(1)) & \mu(1) & \lambda(1) & 0 \\ \vdots & -(\lambda(1)+\mu(2)) & 0 & \lambda(1) \\ & & -(\lambda(2)+\mu(1)) & \mu(1) \\ 0 & \cdots & & -(\lambda(2)+\mu(2)) \end{bmatrix}$$

Fig. 2.15 PHD of the minimum of two generalized Erlang PHDs. (**a**) Diagrammatic representation of the PHD of the minimum. (**b**) The sub generator $\mathbf{D}_0^{(C)}$, in particular $\boldsymbol{\pi}^{(C)} = [\beta\gamma, \beta(1-\gamma), (1-\beta)\gamma, (1-\beta)(1-\gamma)$

Example 2.6. Let us consider concrete examples of the last two operations. We obtain the random variable $X^{(C)}$ which is the minimum of two generalized Erlang distributed random variables $X^{(A)}$, $X^{(B)}$. The representation of the corresponding PHDs is given by

$$\boldsymbol{\pi}^{(A)} = [\beta, 1-\beta],\ \mathbf{D}_0^{(A)} = \begin{bmatrix} -\lambda(1) & \lambda(1) \\ 0 & -\lambda(2) \end{bmatrix}, \qquad \boldsymbol{\pi}^{(B)} = [\gamma, 1-\gamma],\ \mathbf{D}_0^{(B)} = \begin{bmatrix} -\mu(1) & \mu(1) \\ 0 & -\mu(2) \end{bmatrix}.$$

The transition rate matrix of the expanded process representing the distribution of the minimum is given in Fig. 2.15.

Now the case where the random variable $X^{(D)}$ is the maximum of both defined random variables $X^{(A)}$, $X^{(B)}$ is treated. The PHD of the process describing the $max(X^{(A)}, X^{(B)})$ is given in Fig. 2.16. Note that the submatrix $\mathbf{D}_0^{(C)}$ of $\mathbf{D}_0^{(D)}$ is defined in Fig. 2.15b. In particular, the initial distribution vector is given by $\boldsymbol{\pi}^{(D)} = [\beta\gamma, \beta(1-\gamma), (1-\beta)\gamma, (1-\beta)(1-\gamma), 0, 0, 0, 0]$.

2.5 Concluding Remarks

PHDs have a long history dating back to the early work of Erlang [53]. Since then numerous papers have appeared on the subject such that any introduction of the topic must be incomplete which has already been mentioned by Neuts who gave an introduction of PHDs in his famous book on matrix geometric solutions [125]. Like Neuts we tried to introduce PHDs in a form that allows one to use them in computational methods. Later, in Sect. 6, we present models where PHDs are used as building blocks to model event times, and we also outline how these models are analyzed numerically or by simulation. The representation of PHDs using matrices and vectors is also useful for computational algorithms for the parameterization of PHDs which are presented in the following chapter.

a

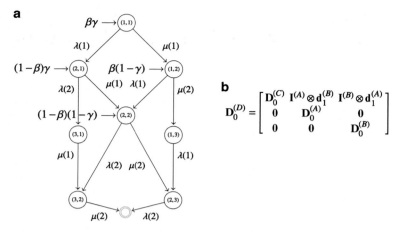

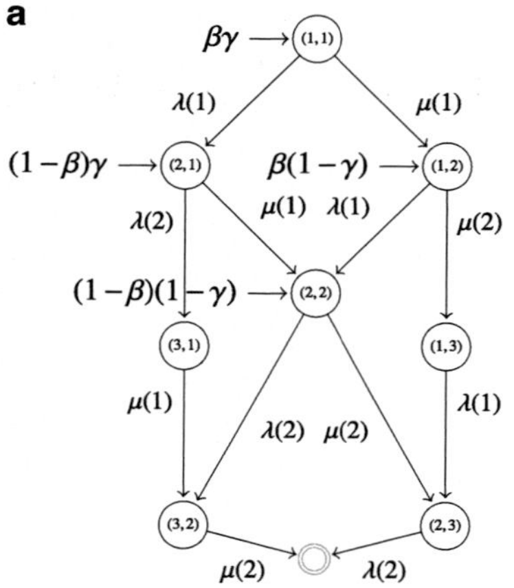

b

$$\mathbf{D}_0^{(D)} = \begin{bmatrix} \mathbf{D}_0^{(C)} & \mathbf{I}^{(A)} \otimes \mathbf{d}_1^{(B)} & \mathbf{I}^{(B)} \otimes \mathbf{d}_1^{(A)} \\ \mathbf{0} & \mathbf{D}_0^{(A)} & \mathbf{0} \\ \mathbf{0} & \mathbf{0} & \mathbf{D}_0^{(B)} \end{bmatrix}$$

Fig. 2.16 PHD of the maximum of two generalized Erlang PHDs. (**a**) Diagrammatic representation of the PHD of the maximum. (**b**) The sub generator $\mathbf{D}_0^{(D)}$

Several additional aspects of PHDs are handled in the literature. The eigenvalues of the matrix $\mathbf{D}_0$ and the Laplace transform play an important role in the characterization of PHDs and in the finding of canonical and minimal representations. Fundamental work of Cox [45] and many successors [66, 121, 129, 131] present results in this directions. However, a canonical representation for general PHDs is missing and also the question whether a given PHD has a representation with less states is still unanswered.

We consider here only PHDs in continuous time. As already mentioned, it is obvious that also discrete time PHDs based on DTMCs rather than CTMCs can be defined. Several results can be transferred from the continuous time to the discrete time area, but there are also some specific aspects that need to be considered, details can be found in [20, 103]. Additionally, one can neglect the probabilistic interpretation and interpret the vector matrix pair describing a PHD in a linear algebraic context. The resulting distributions are denoted as matrix exponential distributions [54, 67, 111] but will not be considered here since the corresponding theory and the practical applicability is less advanced.

Chapter 3
Parameter Fitting for Phase Type Distributions

An important step when developing models that will be subject to a numerical or simulative analysis is the definition of input data for e.g. inter-event or service times which is denoted as input modeling. Usually, one has some observations measured in a real system, called trace, and tries to estimate (fit) the parameters of a distribution, such that the distribution captures characteristics of the given data. In this book we consider Markov processes as models for the data. As already mentioned in Chap. 2 the matrix representation of a PHD is highly redundant in general, which makes fitting of PHDs a difficult optimization problem. Optimization algorithms have to deal with more parameters than necessary and might run from one representation into another equivalent representation of the PHD. As we will see throughout this chapter, many of the existing fitting approaches for PHDs are tailored to specific subclasses for which canonical representations exist (cf. Chap. 2) to avoid problems with redundant representations of the PHDs.

In general, fitting approaches for PHDs can be divided in two classes, depending on the information from the trace they use. The first class of techniques, usually Expectation Maximization (EM) algorithms, uses the complete trace for parameter estimation, while the second class only uses some derived measures like moments.

In the following we will first introduce traces and measures derived from them more formally. After that we will present some approaches that use the complete information available from the trace. Then an overview of algorithms that use derived measures like moments for parameter estimation is given.

3.1 Trace Based Fitting

The objective of a fitting procedure is the computation of a PHD with a representation $(\boldsymbol{\pi}, \mathbf{D}_0)$ with statistical properties equal or similar to the properties of the process of interest $\mathcal{P}$. Since the behavior of process $\mathcal{P}$ from the real system or from an adequate simulation model cannot be infinitely observed, it is usually substituted

P. Buchholz et al., *Input Modeling with Phase-Type Distributions and Markov Models: Theory and Applications*, SpringerBriefs in Mathematics,
DOI 10.1007/978-3-319-06674-5_3, © Peter Buchholz, Jan Kriege, Iryna Felko 2014

by a finite observed sequence of data, e.g. inter-arrival or service times, which is denoted as a trace $\mathcal{T} = (t_1, \ldots, t_m)$. The trace $\mathcal{T}$ is said to be a realization of the process $\mathcal{P}$, thus its statistical properties should resemble the characteristics of the underlying process $\mathcal{P}$. A statistical measure $\xi \in \Xi$, where Ξ is a set of statistical measures characterizing process $\mathcal{P}$, can be either directly given by $\xi_{\mathcal{P}}$ or can be estimated from the trace $\mathcal{T}$ as $\xi_{\mathcal{T}}$. For a given $\xi_{\mathcal{P}}$ or $\xi_{\mathcal{T}}$ the fitting procedure approximates this value by the statistical measure $\xi_{(\boldsymbol{\pi},\mathbf{D}_0)}$ computed from a PHD with a representation $(\boldsymbol{\pi}, \mathbf{D}_0)$. In the following we briefly describe statistical measures which could be obtained from traces.

3.1.1 Definition of Traces and Derived Measures

When faced with an applied probability problem like generation of appropriate input models, an obvious approach is to exploit preliminary observations of processes and to collect the data into traces, which contain the measured data points in a form applicable for further analysis. A trace $\mathcal{T}$ is defined as a sequence of m chronologically ordered points in time $t_i > 0$, $i = 1, \ldots, m$. Traces result from measurements of real systems and can be derived automatically or semi-automatically during the operation of the system using appropriate software which is the case in many application areas, like computer networks. For example the measurements of the input process to a router in a computer network requires bookkeeping of the time steps of incoming packets. In other areas, like for example, the analysis of failure times, measurements require some manual support and cannot be fully automated. One has to be aware that any measurement might be biased, e.g. because the resolution of the system clock is too coarse, or contain errors or outliers. These problems are beyond the scope of this book but one should be aware that every trace first has to be carefully analyzed and should not be used blindly. In the Internet repositories a large number of example traces are available, including measurements for traffic in computer networks [84], failure times and availability data [44, 56], vehicular traffic [142, 159], and many other application areas.

Usually, an element t_i describes the inter-event time of the ith event, but a trace could as well contain service times, packet sizes or other data, depending on the application area. The sequence $t_1, \ldots, t_m$ is assumed to be in a strict-sense stationary, which implies a common distribution of t_i independent on i. For parameter fitting traces can either be used directly or first some measures are estimated from the trace and the parameters of a PHD are afterwards set to approximate the derived measures as close as possible. In general both classes of fitting approaches, using the complete trace or derived measures, have their advantages and disadvantages. Since traces are often very large, containing a million or more measurements, trace based fitting requires often much time but exploits the complete available information, whereas fitting methods based on derived measures are usually much more efficient but use only information in a condensed form. In the subsequent sections methods of both types are presented. We first define some basic quantities characterizing a trace that are later used in algorithms for parameter fitting.

The estimator for the i th moment of the trace and the variance are given by

$$\hat{\mu}_i = \frac{1}{m}\sum_{j=1}^{m}(t_j)^i \quad \text{and} \quad \hat{\sigma}^2 = \frac{1}{m-1}\sum_{j=1}^{m}(t_j - \hat{\mu}_1)^2. \tag{3.1}$$

If the trace exhibits dependencies between consecutive data points, the autocorrelation or the joint moments are of interest. The coefficient of autocorrelation of data points that are lag k apart is estimated by

$$\hat{\rho}_k = \frac{1}{(m-k-1)\hat{\sigma}^2}\sum_{j=1}^{m-k}(t_j - \hat{\mu}_1)(t_{j+k} - \hat{\mu}_1). \tag{3.2}$$

The estimator of the joint moments $\mu_{i,j} = E[X_k^i\, X_{k+1}^j]$ of two consecutive data points is given by

$$\hat{\mu}_{ij} = \frac{1}{m-1}\sum_{j=1}^{m-1}(t_k)^i(t_{k+1})^j. \tag{3.3}$$

The empirical distribution function of a trace is given by a step function with m steps:

$$F_{\mathcal{T}}(x) = \frac{\sum_{j=1}^{m}\delta(t_j \le x)}{m}, \qquad \delta(b) = \begin{cases} 1 & \text{if } b = \mathit{true}, \\ 0 & \text{if } b = \mathit{false}. \end{cases} \tag{3.4}$$

As already mentioned traces might contain a huge number of measurements such that parameter fitting becomes inefficient when applied to the whole trace. To still use fitting algorithms of the expectation maximization type, as presented in the following sections, it is possible to consider weighted intervals or the number of elements in an interval rather than all detailed values. The resulting trace can be denoted as an aggregated trace. Aggregated traces may be generated from detailed traces, that include exact inter-event times, by so called *trace aggregation* or they may naturally result from measurements with a limited resolution where inter-event times can only be determined up to some precision.

In [139] a trace aggregation method has been proposed which allows one to represent a trace with detailed measurements by an aggregated trace. The empirical distribution function of the trace can be separated into a small number of intervals. Let $\Delta_0 < \Delta_1 < \ldots < \Delta_M$ be the boundaries of the intervals. Then, the trace elements in each interval can be represented by their mean values and the weight of each portion to the entire data trace. More formally, let $\mathcal{T}^* = \{(\hat{t}_1, w(\hat{t}_1)), \ldots, (\hat{t}_{m^*}, w(\hat{t}_{m^*}))\}$ be the aggregated data trace, where $\hat{t}_i$ is the mean value of the trace elements in the interval $(\Delta_{i-1},\Delta_i]$, and w_i is the portion of the elements which fall in that interval. In the uniform aggregation method, the data trace is partitioned into M intervals of identical width such that the values $\hat{t}_i$, w_i can be computed for each interval. Let $min(\mathcal{T})$ be the smallest scale and $max(\mathcal{T})$ be the largest scale that should be considered for the trace aggregation. Then the interval $[min(\mathcal{T}), max(\mathcal{T})]$ is partitioned into M subintervals $(\Delta_{i-1},\Delta_i]$, where $i = 1,\ldots,M$, $0 \le \Delta_0 < min(\mathcal{T})$, $\Delta_M = max(\mathcal{T})$, and the interval length is determined by the selected parameter M, i.e. $\Delta = \Delta_i - \Delta_{i-1} = (max(\mathcal{T}) - \Delta_0)/M$. Let J_i be the set of indices such that $j \in J_i$ if $t_j \in (\Delta_{i-1},\Delta_i]$ for $i = 1,\ldots,M$, and $m_i = |J_i|$ the number of the trace elements in the set J_i. The mean value of the trace elements in an interval $(\Delta_{i-1},\Delta_i]$ can be computed as $\hat{t}_i = \frac{1}{m_i}\sum_{j\in J_i} t_j$. The weight w_i is given by $\frac{m_i}{m}$, and equals the probability $P(j \in J_i) = P(t_j \in (\Delta_{i-1},\Delta_i])$. Note, that intervals with zero mean, i.e. $\hat{t}_i = 0$, can be ignored, thus the number of data elements in the uniformly aggregated trace is $m^* \le M$.

The method has the disadvantage that for a small number of intervals M bad approximations of the main part of the distribution can arise, since a few intervals contain too many values, others may contain only a few trace elements, for example for heavy-tailed distributions. In that case, a logarithmic trace aggregation can be applied, where intervals are chosen with equidistant width on a logarithmic scale, e.g. $(10^{-2},10^{-1}],(10^{-1},10^{0}],(10^{0},10^{1}],(10^{1},10^{2}]$ with the scale parameter $s = 1$. Let $s_{min} = \lfloor \log_{10} min(\mathcal{T}) \rfloor$ be the smallest logarithmic scale, and $s_{max} = \lceil \log_{10} max(\mathcal{T}) \rceil$ be the largest logarithmic scale for the aggregation. The logarithmic interval $(10^s, 10^{s+1}]$ can be further divided using the uniform trace aggregation method with parameter r. Then, for $s = s_{min},\ldots,s_{max}-1$ the interval $(10^s,10^{s+1}]$ is divided into r subintervals $(\Delta_0,\Delta_1],\ldots,(\Delta_{r-1},\Delta_r]$, with $\Delta_i = 10^{(s+\frac{i}{r})}$, $i = 0,\ldots,r$. For example using parameters $s_{min} = -2, s_{max} = 2, r = 5$ the interval $(10^{-1},10^{0}]$ is divided into five subintervals $(10^{-1},10^{(-1+\frac{1}{5})}],(10^{(-1+\frac{1}{5})},10^{(-1+\frac{2}{5})}],(10^{(-1+\frac{2}{5})},10^{(-1+\frac{3}{5})}]$, $(10^{(-1+\frac{3}{5})},10^{(-1+\frac{4}{5})}],(10^{(-1+\frac{4}{5})},10^{0}]$. The logarithmic intervals with higher s index can capture the behavior of the heavy-tailed distributions accurately since they contain enough large data values to describe the tail adequately. The number of elements in the aggregated trace is $m^* \le r(s_{max} - s_{min})$ for the logarithmic aggregation.

The use of trace aggregation before using an Expectation Maximization algorithm (see Sects. 3.1.2–3.1.4) for parameter fitting is recommended for long traces with many similar elements.

If the detailed inter-event times are not available, then the mean values $\hat{t}_i$ cannot be computed. In this case, we consider the trace format with $\Delta_i < \Delta_{i+1}$

$$\bar{\mathcal{T}} = (((\Delta_0,\Delta_1],m_1),((\Delta_1,\Delta_2],m_2),\ldots,(\Delta_{M-1},\Delta_M),m_M)$$

which includes all available information, namely the boundaries of the intervals and the number of elements in the intervals.

In the representations $\mathcal{T}^*$ and $\bar{\mathcal{T}}$ the order of the elements in the trace is not preserved, only the percentage or number of elements with an inter-event time in a given interval is available. This implies that the traces cannot be used to describe dependencies between inter-event times which are available in the original trace $\mathcal{T}$. To describe the dependencies in compact form or consider dependencies in traces derived from measurements with a limited resolution, we define grouped traces $\tilde{\mathcal{T}}$. Let T be the length of the observation period of the whole trace and let $T_0 = 0 < T_1 < \ldots < T_M = T$ be time points that define observation intervals $(T_{i-1}, T_i]$. Furthermore, let m_i be the number of events that are observed in the ith interval, then $\tilde{\mathcal{T}} = (((T_0, T_1], m_1), \ldots, ((T_{M-1}, T_M], m_M))$ is a grouped trace. The intervals often have the same length such that $T = M * \Delta$ and the ith interval equals $((i-1)\Delta, i\Delta]$. A trace in format $\tilde{\mathcal{T}}$ cannot be transformed in any of the other formats without introducing an additional approximation.

3.1.2 Expectation Maximization Approach for General PH Distributions

The Expectation Maximization (EM) algorithm is an iterative method for finding the maximum-likelihood estimate of parameters of an underlying distribution from given trace data [15,117]. The data can be thought as a result of a larger unobserved or only partially observed experiment and thus is incomplete or has missing values [51,90]. A maximum-likelihood approach for APHDs in canonical form was first proposed in [18]. The first EM algorithm for general PHDs was developed by Asmussen et al. [6]. Later extensions of the approach apply uniformization to the continuous time Markov chain [151] and result in more efficient and stable realizations of the EM algorithm [31,35,97]. After some basic explanations we will summarize the EM approach for general PHDs and present the improvement using uniformization. At the end of the section we briefly show how the EM algorithm can be applied for grouped and truncated data. In the following subsections EM algorithms tailored to specific subclasses of PHDs are presented.

Suppose that a trace $\mathcal{T} = (t_1, \ldots, t_m)$ with the density function f is observed where $f(t_i|\mathbf{\Theta})$ is defined as the density function of a single data element t_i with a set of free parameters of a distribution $\mathbf{\Theta}$. More precisely, we have incomplete data $\mathcal{T}$ that is observed or generated by some distribution, which represents only partial observations of a larger unobserved experiment resulting in a whole data X. Thus, there is a many-to-one mapping $\mathcal{T} = u(X)$. Then, the tuple $z = (X, \mathcal{T})$ determines the complete data. Denote by $f(X|\mathcal{T}, \mathbf{\Theta})$ the conditional density of X given $\mathcal{T} = u(X)$, then the joint density function can be specified as $f(z|\mathbf{\Theta}) = f(X, \mathcal{T}|\mathbf{\Theta}) = f(X|\mathcal{T}, \mathbf{\Theta}) f(\mathcal{T}|\mathbf{\Theta})$.

Assume now that the trace data entries t_i are independent and identically distributed with density f. Then the resulting density of the trace, which also defines the likelihood function according to the parameters of the distribution, can be written as

$$\mathcal{L}(\mathbf{\Theta}|\mathcal{T}) = f(\mathcal{T}|\mathbf{\Theta}) = \prod_{i=1}^{m} f(t_i|\mathbf{\Theta}). \tag{3.5}$$

Function $\mathcal{L}(\boldsymbol{\Theta}|\mathcal{T})$ determines the likelihood of the observation $\mathcal{T}$ from a model with parameters $\boldsymbol{\Theta}$. In many cases the log-likelihood $\log(\mathcal{L}(\boldsymbol{\Theta}|\mathcal{T})) = \log(\prod_{i=1}^{m} f(t_i|\boldsymbol{\Theta})) = \sum_{i=1}^{m} \log f(t_i|\boldsymbol{\Theta})$ is used instead of Eq. (3.5), because the logarithmic transformation does not alter the location of the extremum and the sum is easier to handle than the product in optimization algorithms.

If Ω is the space of model parameters, the maximum-likelihood problem is to find a set of model parameters $\boldsymbol{\Theta}^*$ that maximizes the likelihood $\mathcal{L}$, i.e. one is interested in

$$\boldsymbol{\Theta}^* = \arg\max_{\boldsymbol{\Theta}\in\Omega} \mathcal{L}(\boldsymbol{\Theta}|\mathcal{T}). \tag{3.6}$$

The EM algorithm is an iterative method that works in two steps, an expectation (E) and a maximization (M) step. The $(r+1)$th step in the EM algorithm consists of finding a value $\boldsymbol{\Theta}^{r+1}$ which maximizes

$$E(\log f(X,\mathcal{T}|\boldsymbol{\Theta}^{r+1})\,|\,f(X|\mathcal{T},\boldsymbol{\Theta}^{r})) \tag{3.7}$$

where $\mathcal{T}$ is the observed data and $\boldsymbol{\Theta}^r$ is the current estimate of the parameters after r steps of the EM algorithm.

The evaluation of the conditional expectation of the complete data likelihood $f(X,\mathcal{T}|\boldsymbol{\Theta})$ is the E step. The E step finds the distribution of the unobserved data, given the known values for the observed variables and the current estimate of the distribution parameters $\boldsymbol{\Theta}^r$. The M-step reestimates the parameters to be those with the maximum-likelihood, under the assumption that the distribution found in the E step is correct. The EM algorithm is a local maximization algorithm which generates a sequence of estimates with a non-decreasing likelihood, but the sequence may result in a local maximum or even a saddle point [51, 160].

For PHDs the observed data consists of the absorption times which correspond to the trace data. The unobserved data corresponds to the states of the PHD that are visited prior to absorption. The fitting problem is then to find $\boldsymbol{\Theta} = (\boldsymbol{\pi}, \mathbf{D}_0)$ such that the density of the trace values is maximized. The likelihood function is then defined as

$$\mathcal{L}((\boldsymbol{\pi},\mathbf{D}_0)|\mathcal{T}) = \prod_{i=1}^{m} \boldsymbol{\pi} e^{\mathbf{D}_0 t_i}\mathbf{d}_1, \tag{3.8}$$

where the value $\mathcal{L}((\boldsymbol{\pi},\mathbf{D}_0)|\mathcal{T})$ gives a likelihood that the trace $\mathcal{T}$ is generated from the PHD with representation $(\boldsymbol{\pi},\mathbf{D}_0)$. The optimization goal is the maximization of the likelihood $\mathcal{L}((\boldsymbol{\pi},\mathbf{D}_0)|\mathcal{T})$, i.e. Eq. (3.6) becomes

$$(\boldsymbol{\pi},\mathbf{D}_0)^* = \arg\max_{(\boldsymbol{\pi},\mathbf{D}_0)} \prod_{i=1}^{m} \boldsymbol{\pi} e^{\mathbf{D}_0 t_i}\mathbf{d}_1. \tag{3.9}$$

Observation $t_i \in \mathcal{T}$, $i = 1,\ldots,m$, of the time till absorption are the incomplete observation of the Markov process $\{X(t)\}_{0\le t<t_i}$ (cf. Sect. 2.1.1). Though, t_i is the ith realization of a PH distributed random variable Y describing the time till absorption in the Markov process $\{X(t)\}_{0\le t<t_i}$. Since only the value t_i is known, the background Markov process remains unobserved in the sense that there are no information how the Markov process entered an absorbing phase, where it started, which of the phases have been visited and for how long it stayed in each of the visited phases. The complete data can be constructed from the embedded Markov process $\{X_r\}_{0\le r<t_i}$ (cf. Sect. 2.1.1) using the sequence of the visited phases and the sequence of the corresponding phase holding times. Denote by k the number of steps in the Markov process $\{X_r\}_{0\le r<t_i}$ until the absorption occurs, thus $X_k = n+1$, i.e. is the absorbing state of a n-order PHD. The parameter n, defining the dimension of the PHD, is predefined before the optimization starts. A larger n usually results in a larger likelihood but also increases the effort because the number of parameters is in $O(n^2)$ and since a PHD is determined by only $2n-1$ parameters, the representation becomes also more redundant with a larger n. The sequence of the visited phases till absorption can be denoted as $X_0,\ldots,X_{k-1}$, and the sequence of the corresponding holding times as $S_0,\ldots,S_{k-1}$. Then the tuple $z = (x_0,\ldots,x_{k-1},s_0,\ldots,s_{k-1})$ describes the complete behavior of the embedded Markov processes $\{X_r\}_{0\le r<t_i}$ until absorption, i.e. on the interval $[0,t_i]$. Thus the phases' holding times must satisfy $t_i = s_0+\ldots+s_k$. The density $f(z|\boldsymbol{\Theta})$ of the complete observation z equals

$$
\begin{aligned}
f(z|(\boldsymbol{\pi},\mathbf{D}_0)) &= \boldsymbol{\pi}(x_0)\lambda(x_0)e^{-\lambda(x_0)s_0}\mathbf{P}(x_0,x_1)\cdots\lambda(x_{k-1})e^{-\lambda(x_{k-1})s_{k-1}}\mathbf{P}(x_{k-1},n+1)\\
&= \boldsymbol{\pi}(x_0)e^{-\lambda(x_0)s_0}\mathbf{D}_0(x_0,x_1)\cdots e^{-\lambda(x_{k-1})s_{k-1}}\mathbf{d}_1(x_{k-1}).
\end{aligned}
$$

The trace data $\mathcal{T} = \{t_1,\ldots,t_m\}$ includes the outcomes of m independent replications of the Markov process $\{X(t)\}_{0\le t<t_i}$, where $t_i \in \mathcal{T}$. Hence, the observation z describes m versions of visited phases and state holding times till absorption $z = (x_0^1,\ldots,x_0^m,\ldots,s_{k-1}^1,\ldots,s_{k-1}^m)$. Then the density of the observation z, which is also used as likelihood function to be optimized in the EM algorithm, can be written in the form

$$
\mathcal{L}((\boldsymbol{\pi},\mathbf{D}_0)|\mathcal{T}) = f(z|(\boldsymbol{\pi},\mathbf{D}_0)) = \prod_{i=1}^{n}\boldsymbol{\pi}(i)^{B_i}\prod_{i=1}^{n}e^{Z_i\mathbf{D}_0(i,i)}\prod_{i=1}^{n}\prod_{j=1,j\neq i}^{n+1}\mathbf{D}_0(i,j)^{N_{ij}},
$$

where the number of times the Markov process started in phase i is denoted by B_i, the total time spent in phase i is given by Z_i, and the total observed number of jumps from phase i to phase j is N_{ij}, for $i \neq j$, $i \in \mathcal{S}_T$, and $j \in \mathcal{S}$.

For the computation of the conditional expectations we define the following vectors and matrix.

$$\mathbf{f}_{(\boldsymbol{\pi},\mathbf{D}_0),t} = \boldsymbol{\pi} e^{\mathbf{D}_0 t},\ \mathbf{b}_{(\boldsymbol{\pi},\mathbf{D}_0),t} = e^{\mathbf{D}_0 t}\mathbf{d}_1 \text{ and } \mathbf{F}_{(\boldsymbol{\pi},\mathbf{D}_0),t} = \int_0^t \left(\mathbf{f}_{(\boldsymbol{\pi},\mathbf{D}_0),t-u}\right)^T \left(\mathbf{b}_{(\boldsymbol{\pi},\mathbf{D}_0),u}\right)^T du. \tag{3.10}$$

We denote $\mathbf{f}_.$ as the forward vector, $\mathbf{b}_.$ as the backward vector and $\mathbf{F}_.$ as the flow matrix. Equation 3.10 can be computed with a standard solver for differential equations, like the Runge–Kutta method [7], or using uniformization [151]. For uniformization define $\alpha \geq \max_{i\in\mathcal{S}_T}(|\mathbf{D}_0(i,i)|)$ and let $\beta(\alpha t, k) = e^{-\alpha t}(\alpha t)^k/k!$ be the probability that a Poisson process with parameter αt performs k jumps. Poisson probabilities can be computed efficiently and in a numerically stable way up to machine precision [60]. Furthermore let $\mathbf{P}_0 = \mathbf{D}_0/\alpha + \mathbf{I}$. The value of α is usually chosen slightly larger than the absolute value of the minimal diagonal entry of $\mathbf{D}_0$. Then define for a PHD $(\boldsymbol{\pi}, \mathbf{D}_0)$

$$\begin{aligned} \mathbf{v}^{(0)} &= \boldsymbol{\pi} \text{ and } \mathbf{v}^{(k+1)} = \mathbf{v}^{(k)}\mathbf{P}_0 \\ \mathbf{w}^{(0)} &= \mathbf{d}_1 \text{ and } \mathbf{w}^{(k+1)} = \mathbf{P}_0\mathbf{w}^{(k)} \end{aligned} \tag{3.11}$$

for $k = 0, 1, 2, \ldots$. Then

$$\begin{aligned} \mathbf{f}_{(\boldsymbol{\pi},\mathbf{D}_0),t} &= \sum_{k=0}^{\infty} \beta(\alpha t, k)\mathbf{v}^{(k)}, \quad \mathbf{b}_{(\boldsymbol{\pi},\mathbf{D}_0),t} = \sum_{k=0}^{\infty} \beta(\alpha t, k)\mathbf{w}^{(k)} \\ \mathbf{F}_{(\boldsymbol{\pi},\mathbf{D}_0),t} &= \frac{1}{\alpha}\sum_{k=0}^{\infty} \beta(\alpha t, k+1)\sum_{l=0}^{k} \left(\mathbf{v}^{(l)}\right)^T \left(\mathbf{w}^{(k-l)}\right)^T. \end{aligned} \tag{3.12}$$

If the summation is stopped after $k_{\max}$ steps, the error is bound by $c \cdot \left(1 - \sum_{k=0}^{k_{\max}} \beta(\alpha t, k)\right)$ where $c = 1$ for $\mathbf{f}_{(\boldsymbol{\pi},\mathbf{D}_0),t}$ and $c = \max_{i\in\mathcal{S}_T}(\mathbf{d}_1(i))$ for $\mathbf{b}_{(\boldsymbol{\pi},\mathbf{D}_0),t}$ and $\mathbf{F}_{(\boldsymbol{\pi},\mathbf{D}_0),t}$. Equation 3.12 includes no negative values and is numerically more stable than Eq. (3.10). Additionally, Eq. (3.12) includes an error bound if the summation is stopped. The vectors $\mathbf{v}^{(k)}$ and $\mathbf{w}^{(k)}$ may be reused if $\mathbf{f}_{(\boldsymbol{\pi},\mathbf{D}_0),t}$ and $\mathbf{b}_{(\boldsymbol{\pi},\mathbf{D}_0),t}$ are computed for different values of t, as it is the case here where the vectors have been computed for all values in a trace. The reuse of the vectors implies that the uniformization based approach is usually much more efficient than the direct solution of the resulting differential equations using methods like Runge–Kutta. Observe that

$$\mathcal{L}((\boldsymbol{\pi}, \mathbf{D}_0)|\mathcal{T}) = \prod_{k=1}^{m} \boldsymbol{\pi}\mathbf{b}_{(\boldsymbol{\pi},\mathbf{D}_0),t_k} \text{ and} \log(\mathcal{L}((\boldsymbol{\pi}, \mathbf{D}_0)|\mathcal{T})) = \sum_{k=1}^{m} \log\left(\boldsymbol{\pi}\mathbf{b}_{(\boldsymbol{\pi},\mathbf{D}_0),t_k}\right) \tag{3.13}$$

which can be easily computed from the vectors.

Now assume that $(\boldsymbol{\pi}, \mathbf{D}_0)$ is the current estimate of the PHD, then the conditional expectations of B_i, Z_i and N_{ij} are given by (see [6])

$$
\begin{aligned}
E_{(\pi,\mathbf{D}_0),\mathcal{T}}[B_i] &= \frac{1}{m}\sum_{k=1}^{m}\frac{\pi(i)\mathbf{b}_{(\pi,\mathbf{D}_0),t_k}(i)}{\pi\mathbf{b}_{(\pi,\mathbf{D}_0),t_k}},\\
E_{(\pi,\mathbf{D}_0),\mathcal{T}}[Z_i] &= \frac{1}{m}\sum_{k=1}^{m}\frac{\mathbf{F}_{(\pi,\mathbf{D}_0),t_k}(i,i)}{\pi\mathbf{b}_{(\pi,\mathbf{D}_0),t_k}},\\
E_{(\pi,\mathbf{D}_0),\mathcal{T}}[N_{ij}] &= \frac{1}{m}\sum_{k=1}^{m}\frac{\mathbf{D}_0(i,j)\mathbf{F}_{(\pi,\mathbf{D}_0),t_k}(i,j)}{\pi\mathbf{b}_{(\pi,\mathbf{D}_0),t_k}},\\
E_{(\pi,\mathbf{D}_0),\mathcal{T}}[N_{in+1}] &= \frac{1}{m}\sum_{k=1}^{m}\frac{\mathbf{d}_1(i)\mathbf{f}_{(\pi,\mathbf{D}_0),t_k}(i)}{\pi\mathbf{b}_{(\pi,\mathbf{D}_0),t_k}}
\end{aligned}
\tag{3.14}
$$

If the trace contains entries with weights, i.e, $\mathcal{T}^* = ((t_1, w(t_1)), \ldots, (t_m, w(t_m)))$, then Eq. (3.14) becomes

$$
\begin{aligned}
E_{(\pi,\mathbf{D}_0),\mathcal{T}^*}[B_i] &= \sum_{k=1}^{m} w(t_k)\frac{\pi(i)\mathbf{b}_{(\pi,\mathbf{D}_0),t_k}(i)}{\pi\mathbf{b}_{(\pi,\mathbf{D}_0),t_k}},\\
E_{(\pi,\mathbf{D}_0),\mathcal{T}^*}[Z_i] &= \sum_{k=1}^{m} w(t_k)\frac{\mathbf{F}_{(\pi,\mathbf{D}_0),t_k}(i,i)}{\pi\mathbf{b}_{(\pi,\mathbf{D}_0),t_k}},\\
E_{(\pi,\mathbf{D}_0),\mathcal{T}^*}[N_{ij}] &= \sum_{k=1}^{m} w(t_k)\frac{\mathbf{D}_0(i,j)\mathbf{F}_{(\pi,\mathbf{D}_0),t_k}(i,j)}{\pi\mathbf{b}_{(\pi,\mathbf{D}_0),t_k}},\\
E_{(\pi,\mathbf{D}_0),\mathcal{T}^*}[N_{in+1}] &= \sum_{k=1}^{m} w(t_k)\frac{\mathbf{d}_1(i)\mathbf{f}_{(\pi,\mathbf{D}_0),t_k}(i)}{\pi\mathbf{b}_{(\pi,\mathbf{D}_0),t_k}}
\end{aligned}
\tag{3.15}
$$

The M-step results in the estimation of new parameters for the PHD in order to maximize the likelihood.

$$
\begin{aligned}
\hat{\pi}(i) &= E_{(\pi,\mathbf{D}_0),\mathcal{T}}[B_i],\ \hat{\mathbf{D}}_0(i,j) = \frac{E_{(\pi,\mathbf{D}_0),\mathcal{T}}[N_{ij}]}{E_{(\pi,\mathbf{D}_0),\mathcal{T}}[Z_i]},\\
\hat{\mathbf{d}}_1(i) &= \frac{E_{(\pi,\mathbf{D}_0),\mathcal{T}^*}[N_{in+1}]}{E_{(\pi,\mathbf{D}_0),\mathcal{T}}[Z_i]},\ \hat{\mathbf{D}}_0(i,i) = -(\hat{\mathbf{d}}_1(i) + \sum_{i \neq j}^{n} \hat{\mathbf{D}}_0(i,j)).
\end{aligned}
\tag{3.16}
$$

With these ingredients the complete EM algorithm can be formulated.

Since the algorithm is a local optimization algorithm, the choice of an appropriate initial PHD is important and determines the runtime and the fitting quality. However, it is not always clear how to choose the initial distribution. It should be noticed that zero elements in $\mathbf{D}_0^{(0)}$ or $\pi^{(0)}$ remain zero due the computation of new values in the E-step (Eq. 3.14). This has the consequence that it is possible to use the EM algorithm to compute PHDs of a specific type, like Coxian, APH,

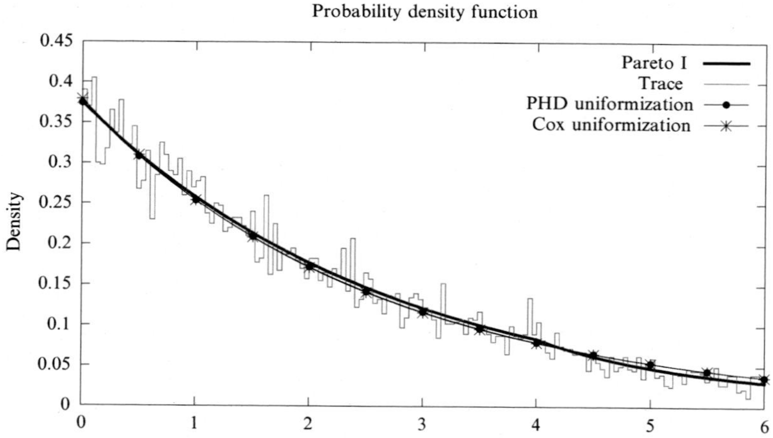

Fig. 3.1 Comparison of the trace, the density of the Pareto distribution and the densities of two fitted PHDs, one with a full matrix and one Coxian distribution

hyper-exponential etc. This also means that an initial representation with many zero entries will result in a PHD with at least the same number of zero entries.

One possibility is to compute an initial PHD using one of the efficient approaches that fit the parameters according to the empirical moments of the trace (see Sect. 3.2). The resulting PHD can then be used as initial distribution for the EM algorithm. However, these approaches usually compute APHDs which means that the resulting PHD will also be acyclic.

Example 3.1. As an example we consider a Pareto I ($\alpha = 1.5, B = 4$) [74] distribution with density $f(t) = \frac{\alpha}{B}e^{-\frac{\alpha}{B}t}$ for $t \leq B$, and $f(t) = \alpha B^{\alpha} e^{-\alpha} t^{-\alpha-1}$ for $t > B$. From this distribution, which will also be used as a benchmark in the subsequent subsections, a trace with 10,000 entries is generated. PHDs with $n = 3$ states are fitted to the trace. The results are shown in Fig. 3.1. It can be seen that both the general PHD and the Coxian distribution provide an excellent fitting quality. The value of the logarithm of the likelihood function is in both cases −21,600, the fitting of the Coxian distribution requires about 12 s on a common PC, the fitting of the PHD takes slightly longer. A larger number of states in the PHD increases the likelihood value only slightly. For $n = 4$ the Coxian distribution reaches a log likelihood value of −21,411, which could not be further improved with $n = 5$. The general PHD requires for $n = 4$ and 5 a huge number of iterations without even reaching the log likelihood value of the PHD with $n = 3$. This shows that without an appropriate initialization, PHDs are initialized randomly here, EM algorithms can be ineffective for a larger number of parameters.

Algorithm 1 determines the parameters of a PHD based on the knowledge of the detailed inter-event times or the mean values in the intervals of an aggregated trace. If the trace $\bar{\mathcal{T}}$ is given, mean values of the inter-event times in an interval are

Algorithm 1 EM algorithm for general PHDs

Input: Trace data $\mathcal{T} = t_1, \ldots, t_m$;
Output: PHD $(\boldsymbol{\pi}, \mathbf{D}_0)$;
1: Choose PHD $(\boldsymbol{\pi}^{(0)}, \mathbf{D}_0^{(0)})$ and set $r = 0$;
2: **repeat**
3: Compute $\mathbf{f}_{(\pi,\mathbf{D}_0),t}$, $\mathbf{b}_{(\pi,\mathbf{D}_0),t}$ and $\mathbf{F}_{(\pi,\mathbf{D}_0),t}$ using $(\boldsymbol{\pi}^{(r)}, \mathbf{D}_0^{(r)})$ in Eq. (3.10) or Eq. (3.12) for $t = t_1, \ldots, t_m$;
4: **E step**: Compute the conditional expectations using Eq. (3.14);
5: **M step**: Compute $(\boldsymbol{\pi}^{(r+1)}, \mathbf{D}_0^{(r+1)})$ using Eq. (3.16) and set $r = r + 1$;
6: **until** $\|\boldsymbol{\pi}^{(r)} - \boldsymbol{\pi}^{(r-1)}\| + \|\mathbf{D}_0^{(r)} - \mathbf{D}_0^{(r-1)}\| < \epsilon$;
7: **return** $(\boldsymbol{\pi}^{(r)}, \mathbf{D}_0^{(r)})$;

not known. It is, of course, possible to substitute intervals by their average values but this introduces an additional approximation. Therefore it is preferable to determine the PHD parameters directly from the interval data without an approximation. This can be done using the EM algorithm presented in [135], which is an extension of the EM algorithm for inter-event times.

We assume that a trace $\bar{\mathcal{T}}$ as defined in Sect. 3.1.1 is available and the parameters of an order n PHD should be fitted to maximize the likelihood function. We assume that the number of events in all intervals are known. The approach can be easily extended to cases where some intervals are not observed [135]. The likelihood function for a PHD $(\boldsymbol{\pi}, \mathbf{D}_0)$ and trace $\bar{\mathcal{T}}$ are then given by

$$\mathcal{L}((\boldsymbol{\pi}, \mathbf{D}_0)|\bar{\mathcal{T}}) = \left(\frac{m!}{\prod_{k=1}^{M} m_k!} \right) \prod_{k=1}^{M} \left(\frac{\int_{\Delta_{k-1}}^{\Delta_k} \boldsymbol{\pi} e^{\mathbf{D}_0 \tau} \mathbf{d}_1 \, d\tau}{\sum_{k=1}^{M} \int_{\Delta_{k-1}}^{\Delta_k} \boldsymbol{\pi} e^{\mathbf{D}_0 \tau} \mathbf{d}_1 \, d\tau} \right)^{m_k} . \tag{3.17}$$

Now consider the following vectors

$$\begin{aligned} \tilde{\mathbf{f}}_{(\pi,\mathbf{D}_0),t} &= \boldsymbol{\pi}(-\mathbf{D}_0)^{-1} e^{\mathbf{D}_0 t}, \ \tilde{\mathbf{b}}_{(\pi,\mathbf{D}_0),t} = e^{\mathbf{D}_0 t} \mathbb{1} \text{ and} \\ \tilde{\mathbf{F}}_{(\pi,\mathbf{D}_0),t} &= \int_0^t \left(\tilde{\mathbf{f}}_{(\pi,\mathbf{D}_0),t-u}\right)^T \left(\tilde{\mathbf{b}}_{(\pi,\mathbf{D}_0),u}\right)^T du. \end{aligned} \tag{3.18}$$

Observe that Eq. (3.18) differs in several details from Eq. (3.10). The vectors can be computed with uniformization using Eq. (3.12) with $\mathbf{v}^{(0)} = \boldsymbol{\pi}(-\mathbf{D}_0)^{-1}$ and $\mathbf{w}^{(0)} = \mathbb{1}$. From the vectors and matrix $\tilde{\mathbf{F}}_{(\pi,\mathbf{D}_0),t}$, the following vectors and matrix can be computed and used afterwards in the E-step.

$$\begin{aligned} &\bar{\mathbf{f}}_{(\pi,\mathbf{D}_0),k} = \tilde{\mathbf{f}}_{(\pi,\mathbf{D}_0),\Delta_{k-1}} - \tilde{\mathbf{f}}_{(\pi,\mathbf{D}_0),\Delta_k}, \ \bar{\mathbf{b}}_{(\pi,\mathbf{D}_0),k} = \tilde{\mathbf{b}}_{(\pi,\mathbf{D}_0),\Delta_{k-1}} - \tilde{\mathbf{b}}_{(\pi,\mathbf{D}_0),\Delta_k} \\ &\text{and } \bar{\mathbf{F}}_{(\pi,\mathbf{D}_0),k} = \tilde{\mathbf{F}}_{(\pi,\mathbf{D}_0),\Delta_{k-1}} - \tilde{\mathbf{F}}_{(\pi,\mathbf{D}_0),\Delta_k} + \left(\bar{\mathbf{f}}_{(\pi,\mathbf{D}_0),k}\right)^T \mathbb{1}^T. \end{aligned} \tag{3.19}$$

The E-step for the grouped trace $\bar{\mathcal{T}}$ and PHD $(\boldsymbol{\pi}, \mathbf{D}_0)$ is then given by

$$E_{(\boldsymbol{\pi},\mathbf{D}_0),\bar{\mathcal{T}}}[B_i] = \frac{1}{m}\sum_{k=1}^{K} m_k \frac{\boldsymbol{\pi}(i)\bar{\mathbf{b}}_{(\boldsymbol{\pi},\mathbf{D}_0),k}(i)}{\boldsymbol{\pi}\bar{\mathbf{b}}_{(\boldsymbol{\pi},\mathbf{D}_0),k}},$$

$$E_{(\boldsymbol{\pi},\mathbf{D}_0),\bar{\mathcal{T}}}[Z_i] = \frac{1}{m}\sum_{k=1}^{K} m_k \frac{\bar{\mathbf{F}}_{(\boldsymbol{\pi},\mathbf{D}_0),k}(i,i)}{\boldsymbol{\pi}\bar{\mathbf{b}}_{(\boldsymbol{\pi},\mathbf{D}_0),k}},$$

$$E_{(\boldsymbol{\pi},\mathbf{D}_0),\bar{\mathcal{T}}}[N_{ij}] = \frac{1}{m}\sum_{k=1}^{K} m_k \frac{\mathbf{D}_0(i,j)\bar{\mathbf{F}}_{(\boldsymbol{\pi},\mathbf{D}_0),k}(i,j)}{\boldsymbol{\pi}\bar{\mathbf{b}}_{(\boldsymbol{\pi},\mathbf{D}_0),k}},$$

$$E_{(\boldsymbol{\pi},\mathbf{D}_0),\bar{\mathcal{T}}}[N_{in+1}] = \frac{1}{m}\sum_{k=1}^{K} m_k \frac{\bar{\mathbf{f}}_{(\boldsymbol{\pi},\mathbf{D}_0),k}(i)\mathbf{d}_1(i)}{\bar{\mathbf{f}}_{(\boldsymbol{\pi},\mathbf{D}_0),k}\mathbf{d}_1} \tag{3.20}$$

The computed values can then be used in the M-step Eq. (3.16) to determine the PHD parameters for the next iteration. Algorithm 1 can then be applied to compute the PHD parameters from the grouped data after substituting Eq. (3.10) by Eq. (3.18) and Eq. (3.14) by Eq. (3.20).

3.1.3 Expectation Maximization Approach for Hyper-Erlang Distributions

In this section, an efficient EM algorithm for fitting of HErDs (cf. Sect. 2.3.1) to a trace developed by Thümmler et al. [156] is presented. The approach extends the fitting procedure of El Abdouni Khayari et al. [94] from hyper-exponential (cf. Sect. 2.3.1) to HErD. Since mixtures of Erlang distributions of unlimited order are theoretically as powerful as acyclic or general PHDs (cf. Theorem 1 from [156]), any probability density function of a non-negative random variable can be approximated arbitrarily close by a HErD.

A HErD is defined as a mixture of K mutually independent Erlang distributions weighted with initial probabilities $\pi(1),\ldots,\pi(K)$, where $\pi(i) \geq 0$, and the sum of the $\pi(i)$ is 1. Let s_i denote the number of phases and $\lambda(i)$ be the rate parameter of the ith Erlang distribution, the density is given by Eq. (2.31).

The mixed-density function can be formally given as

$$f(t_i|\boldsymbol{\Theta}) = \sum_{k=1}^{K} \pi(k) f_k(t_i|\lambda(k)), \tag{3.21}$$

where $f_k(t_i|\lambda(k))$ is the density function parameterized by the rate parameter $\lambda(k)$, thus the vector of parameters is $\boldsymbol{\Theta} = \{\pi(1),\ldots,\pi(K),\lambda(1),\ldots,\lambda(K)\}$. Suppose that a trace $\mathcal{T} = \{t_1,\ldots,t_m\}$ is observed and assume that the trace data t_i are independent and identically distributed with the density in Eq. (3.21). Then the resulting log likelihood value of the trace $\mathcal{T}$ can be written as

$$\log \mathcal{L}(\boldsymbol{\Theta}|\mathcal{T}) = \log \prod_{i=1}^{m} f(t_i|\boldsymbol{\Theta}) = \sum_{i=1}^{m} \log \sum_{k=1}^{K} \pi(k)\, f_k(t_i\,|\,\lambda(k)). \tag{3.22}$$

The log-likelihood function Eq. (3.22) is difficult to optimize because it contains the logarithm of a sum. Consider the unobserved data $z_i \in \{1,\ldots,K\}$, $i = 1,\ldots,m$ which determines the Erlang branch that generated the observed trace data t_i. Then the tuple $z = (z_1,\ldots,z_m)$ describes the unobserved data for the trace $\mathcal{T}$ such that $z_i = k$ if the ith trace element t_i was generated by the kth mixture element f_k. Then the log-likelihood of the complete observation $\mathcal{T}, z$ can be written in the form

$$\log \mathcal{L}(\boldsymbol{\Theta}|\mathcal{T},z) = \sum_{i=1}^{m} \log(\pi(z_i)\, f_{z_i}(t_i\,|\,\lambda(z_i))). \tag{3.23}$$

Since the values z_i are unknown, one can interpret them as random values drawn from the random variable Z. Guessing the appropriate parameters $\hat{\boldsymbol{\Theta}} = \{\hat{\pi}(1),\ldots,\hat{\pi}(K),\hat{\lambda}(1),\ldots,\hat{\lambda}(K)\}$ the mixture components $f_k(t_i\,|\,\hat{\lambda}(k))$ can be easily computed for each k and i by evaluating Eq. (2.31) with appropriate values. Then the probability mass function of the unobserved data z given the observed data $\mathcal{T}$ and the estimates $\hat{\boldsymbol{\Theta}}$ can be derived using Bayes' rule

$$q(z_i\,|\,t_i,\hat{\boldsymbol{\Theta}}) = \frac{q(z_i\,|\,\hat{\boldsymbol{\Theta}}) \cdot f(t_i\,|\,z_i,\hat{\boldsymbol{\Theta}})}{f(t_i\,|\,\hat{\boldsymbol{\Theta}})} = \frac{\hat{\pi}(i) \cdot f_{z_i}(t_i\,|\,\hat{\lambda}(z_i))}{\sum_{k=1}^{K} \hat{\pi}(k)\, f_k(t_i\,|\,\hat{\lambda}(k))}, \tag{3.24}$$

and

$$q(z\,|\,\mathcal{T},\hat{\boldsymbol{\Theta}}) = \prod_{i=1}^{m} q(z_i\,|\,t_i,\hat{\lambda}(z_i)). \tag{3.25}$$

The conditional expectation of the complete data log-likelihood with respect to the unknown random variable Z can be computed, given the observed trace data $\mathcal{T}$ and the current parameter estimates $\hat{\boldsymbol{\Theta}}$

$$E(\log \mathcal{L}(\boldsymbol{\Theta}\,|\,Z,\mathcal{T})|\,\mathcal{T},\hat{\boldsymbol{\Theta}}) = \sum_{z \in \{1,\ldots,K\}^m} \log \mathcal{L}(\boldsymbol{\Theta}\,|\,z,\mathcal{T}) \cdot q(z\,|\,\mathcal{T},\hat{\boldsymbol{\Theta}}). \tag{3.26}$$

Using Eqs. (3.23), (3.25), and by rearranging the sums and the product Eq. (3.26) can be evaluated as

$$\begin{aligned} E(\log \mathcal{L}(\boldsymbol{\Theta}\,|\,Z,\mathcal{T})|\,\mathcal{T},\hat{\boldsymbol{\Theta}}) &= \sum_{z \in \{1,\ldots,K\}^m} \sum_{i=1}^{m} \log(\pi(z_i)\, f_{z_i}(t_i\,|\,\lambda(z_i))) \prod_{i=1}^{m} q(z_i\,|\,t_i,\hat{\lambda}(z_i)) \\ &= \sum_{k=1}^{K}\sum_{i=1}^{m} \log(\pi(k)) \cdot q(k|t_i,\hat{\lambda}(k)) + \sum_{k=1}^{K}\sum_{i=1}^{m} \log(f_k(t_i\,|\,\lambda(k))) \cdot q(k|t_i,\hat{\lambda}(k)). \end{aligned} \tag{3.27}$$

Algorithm 2 EM algorithm for HErDs

Input: Trace data $\mathcal{T} = t_1, \ldots, t_m$;
Output: Optimal parameter vector $\boldsymbol{\Theta} = (\pi(1), \ldots, \pi(K), \lambda(1), \ldots, \lambda(K))$;
1: Choose initial parameter estimates $\hat{\boldsymbol{\Theta}} = (\hat{\pi}(1), \ldots, \hat{\pi}(K), \hat{\lambda}(1), \ldots, \hat{\lambda}(K))$;
2: **repeat**
3: Compute the logarithmic form of the density function given by Eq. (2.31) $f_k(t_i \mid \hat{\lambda}(k)) = \hat{\lambda}(k) e^{(s_k-1)\ln(\hat{\lambda}(k)t_i) - \ln(s_k-1)! - \hat{\lambda}(k)t_i}$ for all $i = 1, \ldots, m$, $k = 1, \ldots, K$;
4: **E step**: Compute the pmf of the unobserved data for $i = 1, \ldots, m$, $k = 1, \ldots, K$ $q(k \mid t_i, \hat{\lambda}(k)) = \frac{\hat{\pi}(k) \cdot f_k(t_i \mid \hat{\lambda}(k))}{\sum_{j=1}^{K} \hat{\pi}(j) f_j(t_i \mid \hat{\lambda}(j))}$;
5: **M step**: Compute $\pi(k)$ and $\lambda(k)$ that maximize the conditional expectation Eq. (3.27) for $i = 1, \ldots, m$ according to Eq. (3.28);
6: set $\hat{\boldsymbol{\Theta}} := \boldsymbol{\Theta}$;
7: **until** convergence reached according to the criterion (a) or (b);

In the M-step of the EM algorithm the conditional expectation computed in the E-step has to be maximized due to the parameters $\hat{\boldsymbol{\Theta}}$. For maximizing Eq. (3.27), the first term containing $\pi(k)$ and the second term containing $\lambda(k)$ can be maximized independently. According to [117, 156] simple closed-form expressions for the parameters of a HErD for a given number of Erlang branches K and a given number of the phases per branch s_k can be derived as

$$\pi(k) = \frac{1}{m} \sum_{i=1}^{m} q(k \mid t_i, \hat{\lambda}(k)) \qquad \lambda(k) = \frac{s_k \sum_{i=1}^{m} q(k \mid t_i, \hat{\lambda}(k))}{\sum_{i=1}^{m} q(k \mid t_i, \hat{\lambda}(k)) t_i}. \tag{3.28}$$

Note, that for the mixture density problem the marginal distribution occurring in computation of the expectation Eq. (3.27) and in the closed-form expressions for the parameters of a HErD, can be computed by Eqs. (3.24), (3.25).

The pseudo-code of the EM algorithm is given by Algorithm 2, where the log-likelihood value is guaranteed to increase in each iteration and the algorithm converges to a local maximum of the likelihood function [117]. Additionally, the following criteria could be used in each iteration to check if the convergence has been reached:

(a) the maximal difference of the values of the parameter vectors of successive iterations;
(b) the relative difference of the log-likelihood values of successive iterations.

The EM algorithm can be stopped if the computed difference (a) or (b) is below some sufficiently small ϵ, e.g. $\epsilon = 10^{-6}$.

Note that a straightforward computation of the Erlang densities

$$f_k(t_i \mid \lambda(k)) = \frac{(\lambda(k)t_i)^{s_k-1}}{(s_k-1)!} \lambda(k) e^{-\lambda(k)t_i}$$

can be numerically difficult for large number of phases, e.g. $s_k > 50$, since large factorials and large power values must be computed. For this reason, the logarithmic form of the density function proposed in step 3 together with precomputed logarithms of the factorial values, i.e. $\ln s! = \sum_{i=1}^{s} \ln i$, can be used.

The log-likelihood value according to Eq. (3.22) can be computed in the E-step (step 4 of the EM algorithm), where the computation of the unobserved data pmf has the complexity $\mathcal{O}(mK)$ when computing numerator and denominator separately. The M step has also complexity $\mathcal{O}(mK)$ thus resulting in the overall complexity of the EM algorithm for HErDs $\mathcal{O}(mK)$ which is independent of the number of phases of the HErD.

A further advantage of the EM algorithm is that the continuous parameter vectors $\boldsymbol{\pi}$ and $\boldsymbol{\lambda}$ of a HErD can be optimized for a predefined setting of the number of Erlang branches K and number of phases of each Erlang branch $s_k, k = 1, \ldots, K$. The objective is to determine a setting of a number of phases $s_1 \ldots, s_K$ that maximizes the log-likelihood. For small number of phases of a HErD n, e.g. $n \leq 10$ and $m \leq 10^6$, all possible settings of the number of branches K and $s_1, \ldots, s_K$ can be enumerated and for each such setting a HErD can be fitted. The best HErD can then be obtained by comparing the fitted distributions according to their log-likelihood values. The optimal HErD has a maximum log-likelihood value. The approach becomes inefficient for large n and m, $n > 10$, $m > 10^7$, since the number of possible settings increases exponentially. In this case some fitting strategies recommended in [156] can be used. For example, if the empirical distribution of the trace has a small coefficient of variation, i.e. $C^2 < 1$, it is recommended to fit the HErD only with 1–3 Erlang branches. For monotonically decreasing empirical distribution functions with large coefficient of variation, i.e. $C^2 > 1$, fitting of the HErD with $n, (n-1)$ or $(n-2)$ Erlang branches should be preferred. A special case where $n = K$ corresponds to the hyper-exponential distribution and could be well used for fitting of heavy-tailed distributions with large coefficient of variation [58].

It is also possible to combine the EM algorithm with moment matching approach described in [87–89, 156], where under some specific conditions the first three empirical moments can be matched exactly by only three of the $2K-1$ free continuous parameters of a HErD. For example, one can first obtain a fitted HErD via the EM algorithm and than adjust the moments using the resulting HErD. Since the log-likelihood value decreases only marginal when the first three moments are matched, the EM algorithm can be reinitialized with the obtained moment matching solution. Generally, the returned vector of estimated continuous parameters strongly depends on the initial values $\hat{\boldsymbol{\Theta}}^0$. Thus, initializing the EM algorithm with a solution matching the first three moments exactly forces the EM algorithm to converge to a local maximum with a better tail fitting [156].

An efficient computational speed up of the EM algorithm can be achieved by taking the aggregated trace $\mathcal{T}^*$ of the size m^* into consideration (see the aggregation techniques described in Sect. 3.1.1). Since the aggregated trace $\mathcal{T}^*$ contains the mean values for each interval $(\Delta_{i-1}, \Delta_i]$, $i = 1, \ldots, m^*$, and the probabilities of finding trace values in the intervals, Eq. (3.28) can be restated as

Table 3.1 Elements of the trace with the length $m = 40$

0.000147605	0.243096	0.820047	1.73186	2.12233	2.76842	23.6486	151.602
0.00721833	0.499803	0.341485	1.33033	2.39616	2.78563	10.7602	103.906
0.00177183	0.629885	1.46946	1.09754	2.46193	3.34915	8.78572	72.1299
0.84895	0.104488	1.29368	1.96754	2.0992	3.79031	10.7661	32.1734
0.196494	0.911078	1.53472	1.85547	2.60535	3.54335	39.0293	12.3034

The trace is summarized from the data trace obtained from the Pareto I ($\alpha = 1.5, B = 4$) [74] distribution with the density defined as $f(t) = \frac{\alpha}{B}e^{-\frac{\alpha}{B}t}$ for $t \leq B$, and $f(t) = \alpha B^{\alpha} e^{-\alpha} t^{-\alpha-1}$ for $t > B$. The moments of the trace are $\hat{\mu}_1 = 12.7477$, $\hat{\mu}_2 = 1066.5976$, $\hat{\mu}_3 = 127316.9371$, the trace variance is $\hat{\sigma}^2 = 927.2733$, $C^2 = 5.706$, $min(\mathcal{T}) = 0.000147605$, and $max(\mathcal{T}) = 151.602$

$$\pi(k) = \sum_{i=1}^{m^*} q(k\,|\,\hat{t}_i, \hat{\lambda}(k))\, w(\hat{t}_i) \qquad \lambda(k) = \frac{s_k \sum_{i=1}^{m^*} q(k\,|\,\hat{t}_i, \hat{\lambda}(k))\, w(\hat{t}_i)}{\sum_{i=1}^{m^*} q(k\,|\,\hat{t}_i, \hat{\lambda}(k))\hat{t}_i\, w(\hat{t}_i)}. \tag{3.29}$$

Further advantage is that the computation of the pmf of the unobserved data z in the Eqs. (3.24) and (3.25) has to be done only for m^* elements from the aggregated trace. The log-likelihood value Eq. (3.22) for the aggregated trace can be computed by

$$\log \mathcal{L}(\mathbf{\Theta}|\mathcal{T}^*) = m \sum_{i=1}^{m^*} w(\hat{t}_i) \log\left(\sum_{k=1}^{K} \pi(k)\, f_k(\hat{t}_i\,|\,\lambda(k)) \right). \tag{3.30}$$

In comparison to the general approach, the complexity of the EM algorithm for HErDs can be reduced from $\mathcal{O}(mK)$ to $\mathcal{O}(m^*K)$ using the trace aggregation technique.

Example 3.2. Considering the data trace shown in Table 3.1 the following fitting results have been obtained using the software Gfit [156] where the above methods are implemented. First, the complete data trace was fitted to a HErD of order $n = 4$ with only one phase per branch, i.e. $n = K$, and the number of iterations is 50. The approximating PHD $(\boldsymbol{\pi}^1, \mathbf{D}_0^1)$ is shown in Fig. 3.3 with the corresponding density shown in Fig. 3.2. The approximated moments are $\mu_1 = 12.7477, \mu_2 = 1060.0554$, $\mu_3 = 146593.1446$, and the approximated variance is $\sigma^2 = 897.5493$, $C^2 = 5.523$.

Then the uniform trace aggregation with 8 intervals $[\Delta_{i-1}, \Delta_i)$, $i = 1, \ldots, 8$, was used to fit the aggregated trace $\mathcal{T}^*$ as described in Sect. 3.1.1. Since $min(\mathcal{T}) = 0.000147605$ and $max(\mathcal{T}) = 151.602$, the smallest scale to be considered is $\Delta_0 = 0$ and the largest scale is $\Delta_8 = 151.602$, the interval length is $\frac{151.602}{8} = 18.95025$. Thus, we obtain following intervals $[0, 18.95025)$, $[18.95025, 37.9005)$, $[37.9005, 56.85075)$, $[56.85075, 75.801)$, $[75.801, 94.75124)$, $[94.75124, 113.7015)$, $[113.7015, 132.6517)$, $[132.6517, 151.602)$ with weights $w(\hat{t}_1) = \frac{34}{40}, w(\hat{t}_2) = \frac{2}{40}, w(\hat{t}_3) = \frac{1}{40}, w(\hat{t}_4) = \frac{0}{40}, w(\hat{t}_5) = \frac{1}{40}, w(\hat{t}_6) = \frac{1}{40}, w(\hat{t}_7) = \frac{0}{40}, w(\hat{t}_8) = \frac{1}{40}$, and the mean values for each interval $\hat{t}_1 = 2.5712, \hat{t}_2 = 27.911, \hat{t}_3 = 39.0293, \hat{t}_5 = 72.1299, \hat{t}_6 = 103.906, \hat{t}_8 = 151.602$ such that the aggregated trace

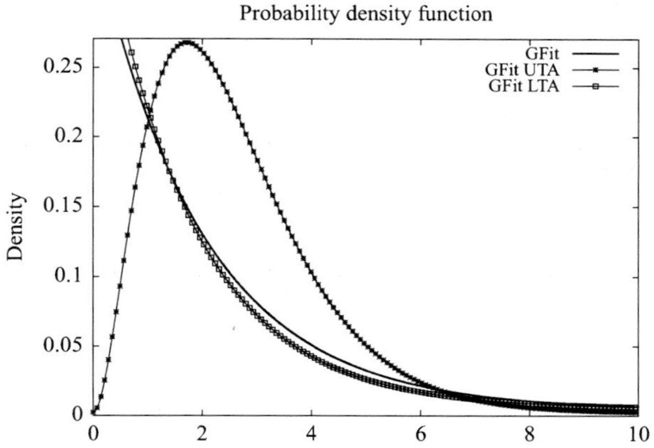

Fig. 3.2 A comparison between the fitting results obtained by the method [139, 156] which is implemented in the tool Gfit. The method is applied for the trace visualized in Table 3.1. The result obtained by Gfit with logarithmic trace aggregation gives an approximation as good as the Gfit result without trace aggregation. The approximation resulting from Gfit with uniform trace aggregation deviates significantly because of the small number of intervals

$$\mathbf{D}_0^1 = \begin{bmatrix} -0.0215 & 0 & 0 & 0 \\ 0 & -0.4885 & 0 & 0 \\ 0 & 0 & -0.5358 & 0 \\ 0 & 0 & 0 & -351.0801 \end{bmatrix}, \pi^1 = [0.2457, 0.4881, 0.1959, 0.0702],$$

$$\mathbf{D}_0^2 = \begin{bmatrix} -0.0149 & 0 & 0 & 0 \\ 0 & -1.1667 & 1.1667 & 0 \\ 0 & 0 & -1.1667 & 1.1667 \\ 0 & 0 & 0 & -1.1667 \end{bmatrix}, \quad \pi^2 = [0.1588, 0.8412, 0, 0],$$

$$\mathbf{D}_0^3 = \begin{bmatrix} -0.0237 & 0 & 0 & 0 \\ 0 & -0.5853 & 0 & 0 \\ 0 & 0 & -0.5854 & 0 \\ 0 & 0 & 0 & -354.0899 \end{bmatrix}, \pi^3 = [0.2759, 0.4846, 0.1697, 0.0696].$$

Fig. 3.3 The PHD with representation $(\boldsymbol{\pi}^1, \mathbf{D}_0^1)$ was fitted using raw Gfit method without trace aggregation. The PHD $(\boldsymbol{\pi}^2, \mathbf{D}_0^2)$ was obtained using uniform aggregation method with 8 intervals, and the PHD $(\boldsymbol{\pi}^3, \mathbf{D}_0^3)$ was fitted using logarithmic aggregation method with equidistant 7 logarithmic intervals and 8 uniform intervals for each logarithmic interval

becomes $\mathcal{T}^* = \{(2.5712, 0.85), (27.911, 0.05), (39.0293, 0.025), (72.1299, 0.025), (103.906, 0.025), (151.602, 0.025)\}$ containing only 6 intervals. Note that mean values $\hat{t}_4, \hat{t}_7$ cannot be determined since the intervals 4 and 7 are empty. The approximating PHD $(\boldsymbol{\pi}^2, \mathbf{D}_0^2)$ with $\mu_1 = 12.7477, \mu_2 = 1418.7567, \mu_3 = 282307.649, \sigma^2 = 1256.2506, C^2 = 7.7304$ is shown in Fig. 3.3.

Finally the logarithmic trace aggregation method with seven equidistant logarithmic intervals was used to obtain a HErD. The logarithmic intervals are $[10^{-4}, 10^{-3})$, $[10^{-3}, 10^{-2}), \ldots$, $[10^1, 10^2)$, $[10^2, 10^3)$ since $min(\mathcal{T}) = 0.000147605 \in [10^{-4}, 10^{-3})$

and $max(\mathcal{T}) = 151.602 \in [10^2, 10^3)$. We divide each logarithmic interval $(10^s, 10^{s+1}]$, $s = s_{min} = 10^{-4}, \ldots, s_{max} - 1 = 10^2$, into $r = 8$ subintervals $(\varDelta_0, \varDelta_1], \ldots, (\varDelta_{r-1}, \varDelta_r]$, with $\varDelta_i = 10^{(s+\frac{i}{r})}$, $i = 0, \ldots, r$. Defining 8 sets of elements for each logarithmic interval as in the uniform trace aggregation we obtain

$$\begin{aligned}\mathcal{T}^* = \{&(0.000147605, 0.025), (0.00177183, 0.025), (0.00721833, 0.025),\\ &(0.104488, 0.025), (0.196494, 0.025), (0.243096, 0.025), (0.341485, 0.025),\\ &(0.499803, 0.025), (0.629885, 0.025), (0.860025, 0.075), (1.2405, 0.075),\\ &(1.5786, 0.075), (2.0111, 0.1), (2.6034, 0.125), (3.5609, 0.075), (8.7857, 0.025),\\ &(11.2765, 0.075), (23.6486, 0.025), (35.6013, 0.05), (72.1299, 0.025),\\ &(103.906, 0.025), (151.602, 0.025)\}\end{aligned}$$

containing 22 elements.

The resulting PHD with representation $(\boldsymbol{\pi}^3, \mathbf{D}_0^3)$ with the approximation closest to the PHD $(\boldsymbol{\pi}^1, \mathbf{D}_0^1)$ obtained without trace aggregation is shown in Fig. 3.3. Its statistics are $\mu_1 = 12.7477, \mu_2 = 983.925, \mu_3 = 123918.3359, \sigma^2 = 821.4189$, and $C^2 = 5.0546$.

3.1.4 *Expectation Maximization Approach for Canonical Representations*

In this section the EM algorithm, which is used to fit the parameters of an APHD in series canonical form (cf. Sect. 2.3) based on the work of Okamura et al. [134], is presented. The approach is an extension of Algorithm 1.

Any APHD of order n admits a minimal representation with only $2n - 1$ free parameters, as shown by Cumani in [47]. The minimal representation is unique, and it is referred to as the series canonical form, since it is defined as a mixture of basic series of an APHD with transition rates in ascending order, i.e. $\lambda(i) \leq \lambda(i+1) \leq \ldots \leq \lambda(n)$ (cf. Definition 2.8). Since EM algorithms keep zero elements, Algorithm 1 can be initialized with an APHD in series canonical form. An iteration of the EM algorithm will preserve the non-zero structure of the matrices which implies that matrix $\mathbf{D}_0$ has non-zero elements in the diagonal and first upper subdiagonal. Furthermore, the last element of $\mathbf{d}_1$ will remain non-zero. However, it cannot be assured that $|\mathbf{D}_0(i,i)| \leq |\mathbf{D}_0(j,j)|$ still holds for $i < j$ after an iteration. This implies that the resulting representation is not necessarily in series canonical form.

Since the EM algorithm cannot guarantee to preserve the parameter constraints $\lambda(1) \leq \lambda(2) \leq \ldots \leq \lambda(n)$, an additional recomputation procedure is required after each EM-step to ensure the canonical ordering of the transition rates. Let $(\boldsymbol{\pi}, \mathbf{D}_0)$ be the current estimate of a PHD of order n which is not in the series canonical form. Then there are some $\lambda(i)$, $\lambda(i+1)$ with $\lambda(i) > \lambda(i+1)$. We denote $\lambda(i)$ as μ and $\lambda(i+1)$ as λ.

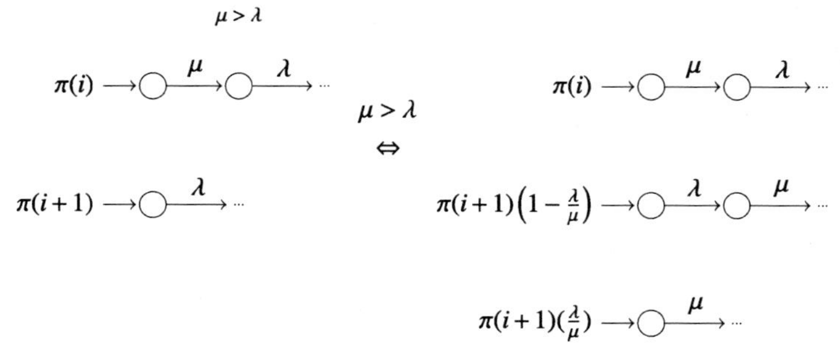

Fig. 3.4 Substitution step for the reconstruction of the series canonical form. The two above elementary series on the right side $<\mu\lambda>$ and $<\lambda\mu>$ are merged into one elementary series $<\lambda\mu>$ with probability $\pi(i)+\pi(i+1)\left(1-\frac{\lambda}{\mu}\right)$

Fig. 3.5 Equivalent representation of the APHD. The APHD shown in Fig. 3.5b has been transformed in the series canonical form since for the APHD in Fig. 3.5a the transition rates are not ordered ascendingly, i.e. $\lambda(i) \nleq \lambda(i+1)$. The 2-phase APHD in Fig. 3.5a has the Laplace transform $F(s) = \frac{2}{3}\frac{5\cdot 3}{(s+5)(s+3)} + \frac{1}{3}\frac{3}{(s+3)} = \frac{s+15}{(s+5)(s+3)}$. The APHD in Fig. 3.5b is cdf-equivalent to the former one since it has the Laplace transform $F(s) = \frac{2}{3}\frac{5\cdot 3}{(s+5)(s+3)} + \frac{1}{3}\left[\frac{2}{5}\frac{3\cdot 5}{(s+3)(s+5)} + \frac{3}{5}\frac{5}{(s+5)}\right] = \left(\frac{2}{3}+\frac{2}{15}\right)\frac{3\cdot 5}{(s+3)(s+5)} + \frac{1}{(s+5)} = \frac{s+15}{(s+5)(s+3)}$

The recomputation can then be done by modifying the parameters $\boldsymbol{\pi}, \mathbf{D}_0$:

$$\lambda(i)^{new} = \lambda,\ \ \lambda(i+1)^{new} = \mu,\ \ \boldsymbol{\pi}(i)^{new} = \boldsymbol{\pi}(i) + \boldsymbol{\pi}(i+1)\left(1-\frac{\lambda}{\mu}\right),$$
$$\boldsymbol{\pi}(i+1)^{new} = \boldsymbol{\pi}(i+1)\left(\frac{\lambda}{\mu}\right). \tag{3.31}$$

We recall that an elementary series with some phase with transition rate λ can be substituted with a mixture of two elementary series, one containing a phase with transition rate $\mu > \lambda$, and the other containing both phases with the rates λ and μ (see substitution step in Fig. 2.12). Hence, the recomputation procedure substitutes the elementary series with transition rate λ with a mixture of two elementary series $<\mu>$ and $<\lambda\mu>$ such that $\lambda < \mu$ and we obtain an ascending ordering of the transition rates. Note that the elementary series $<\mu\lambda>$ already exists which is equivalent to the new elementary series $<\lambda\mu>$. This substitution step is visualized in Fig. 3.4.

Example 3.3. The 2-phase APHD $(\boldsymbol{\pi}, \mathbf{D}_0)$ with a constraint violation results from an EM estimation. In particular, $\boldsymbol{\pi} = \left[\frac{2}{3}, \frac{1}{3}\right]$, $\lambda(1) = 5$, and $\lambda(2) = 3$ such that $\lambda(1) > \lambda(2)$. The APHD is shown in Fig. 3.5. The elementary series with transition

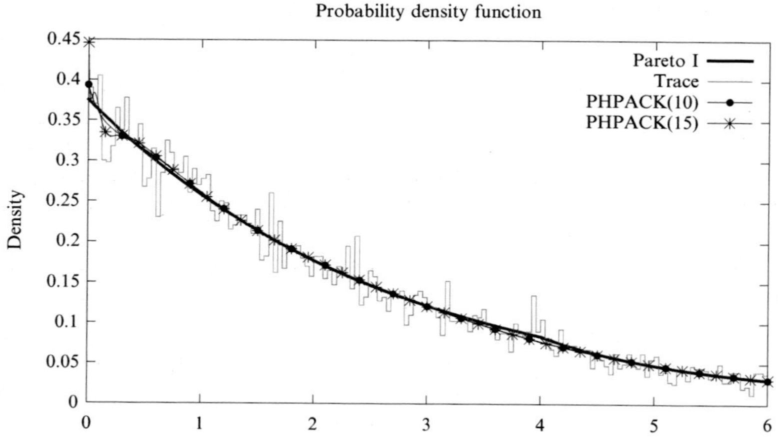

Fig. 3.6 Comparison of the trace, the density of the Pareto distribution and the densities of the two fitted PHDs, one of order 10 denoted as PHPACK(10) and one of order 15 denoted as PHPACK(15)

rate 3 is substituted with a mixture of two elementary series, one containing a phase with transition rate $5 > 3$ which is weighted with a probability $\frac{3}{5}$. The other elementary series $< 35 >$ contains both phases with rates 3 and 5 and is weighted with the remaining probability $\left(1 - \frac{3}{5}\right) = \frac{2}{5}$. Considering the initial probabilities $\frac{1}{3}$ for the elementary series $< 3 >$ we obtain the probability $\frac{1}{3}\frac{3}{5} = \frac{3}{15}$ for the $< 5 >$ and $\frac{1}{3}\left(1 - \frac{3}{5}\right) = \frac{2}{15}$ for the $< 35 >$. Merging the elementary series $< 35 >$ and $< 53 >$ together the probability for $< 35 >$ is $\frac{2}{3} + \frac{2}{15} = \frac{12}{15}$.

The use of the series canonical form for the parameter estimation of APHDs has several advantages compared to the use of general APHDs in the algorithm. First, the effort is proportional to the number of non-zero elements in $\mathbf{D}_0$ and $\mathbf{d}_1$ which is only n for an order n APHD. Secondly, the search area of the EM algorithm is restricted to minimal and unique representations which reduces the number of local maxima and avoids fluctuation between several equivalent representations. The additional effort to transform the APHD after an iteration in series canonical form is usually negligible.

Example 3.4. As an example we again consider a Pareto I ($\alpha = 1.5, B = 4$) [74] distribution with the density defined as $f(t) = \frac{\alpha}{B}e^{-\frac{\alpha}{B}t}$ for $t \leq B$, and $f(t) = \alpha B^{\alpha}e^{-\alpha}t^{-\alpha-1}$ for $t > B$. From this distribution a trace with 10,000 unsorted entries is generated. PHDs with $n = 10$ and $n = 15$ states are fitted to the trace. The results are shown in Fig. 3.6. It can be seen that both resulting PHDs provide an excellent fitting quality. The value of the logarithm of the likelihood function is $-21,386$ in the case of the 10 state PHD, the total fitting procedure requires about 70 s on a common PC. A larger number of states for the PHD increases the likelihood value only slightly such that the PHD of order 15 reaches a log likelihood value of $-21,385$ but requires about 140 s.

3.1.5 Density Based Parameter Fitting

In contrast to the EM algorithms described in preceding sections which minimize the Kullback–Leibler distance, i.e. the cross entropy [6, 58, 74], additional fitting methods minimizing density based distance measures [19], and hybrid techniques have been developed [149, 156]. Particularly, for heavy-tailed distributions it may be possible that maximum likelihood methods cannot capture the tail behavior correctly since the EM algorithm searches for a global maximum [74, 146] and gives more importance to the main part of the density which is represented by the majority of measured elements. Since the tail of any PHD decays exponentially [153], a precise representation of the main part may result in a very bad approximation of the tail behavior. The practical importance of a good approximation of the tail behavior of heavy-tailed distributions has been demonstrated by recent research results, thus fitting of heavy tails needs a special treatment.

In [74] a combined fitting approach is proposed that uses a density based fitting approach for approximating the main part of the original distribution which minimizes an arbitrary distance measure, e.g. density absolute area difference given by $\hat{D} = \int_0^\infty |f(t) - \hat{f}(t)|dt$. Additionally, the heuristic method developed in [58] is used to fit its tail with monotone decreasing density using a hyper-exponential PHD. In the method [74] first the PHD $(\boldsymbol{\pi}, \mathbf{D}_0)^0$ with a minimal distance measure from a set of 200 PHDs having the proper mean $\int_0^\infty tf(t)dt$ of the original distribution $f(t)$ is selected as the initial point from the parameter space. Then the linear programming approach is applied to determine the direction in which the distance measure decreases, and the parameter search proceeds in that direction until the predefined convergence criterion is satisfied.

The recursive method proposed by Feldmann and Whitt in [58] is used to fit completely monotone continuous distributions into hyper-exponential PHDs. The algorithm starts at the largest time scale that should be considered and successively reduces the time scale. First the exponential weighted with the corresponding probability is fitted to the rightmost portion of the tail. In the next recursion step the known weighted exponential can be subtracted from the original distribution, and the second weighted exponential can be fitted to the remaining tail. The heuristic algorithm is very efficient, i.e. a good approximation of hyper-exponential PHDs results in the desired high variability, but the method is only suitable for distribution functions rather than for data traces.

Another approach developed in [146] deals with data traces for fitting monotonically decreasing density functions into hyper-exponential PHDs. The divide-and-conquer method uses a continuous data histogram to divide the data trace into partitions with reduced variability, and fits each partition using the EM algorithm [136]. The final result is obtained by combining hyper-exponential PHDs for all partitions. The procedure is not restricted to completely monotone data traces [58, 146], and can be also applied to traces with not completely monotone empirical distribution functions with one peak only, by constructing a PHD as a mixture of an Erlang and a hyper-exponential PHDs.

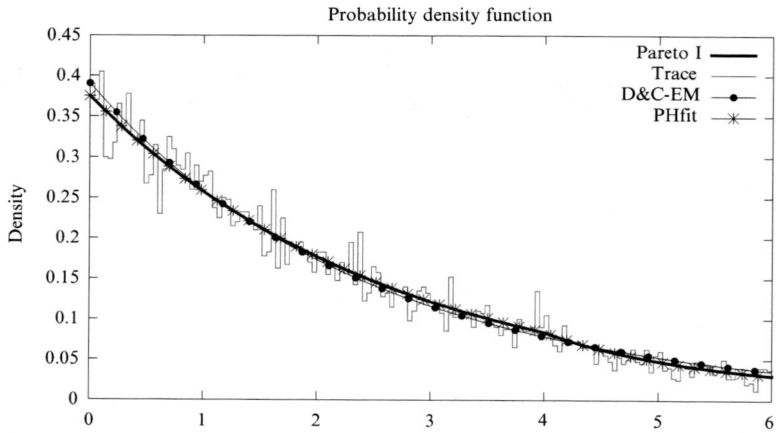

Fig. 3.7 A comparison between the Pareto I ($\alpha = 1.5, B = 4$) [74] density with approximations obtained by the methods [74, 146]. The method [74] is implemented in the tool PHfit, and the method [146] is denoted by D&C-EM which is applied over the trace obtained from the Pareto I ($\alpha = 1.5, B = 4$) [74] distribution. The result of the different emphasis is that the approximation obtained by PHfit approximates the density very accurately such that curves are superposed

Fig. 3.8 The PHD obtained using the D&C-EM method [146] applied to the Pareto trace

$$\mathbf{D}_0 = diag\begin{bmatrix} -0.3807 \\ -0.4142 \\ -0.4696 \\ -0.5303 \\ -0.5895 \\ -0.6530 \\ -0.7349 \\ -0.0326 \\ -0.0407 \\ -0.408 \\ -0.0409 \\ -0.0005 \end{bmatrix} \quad \pi = \begin{bmatrix} 0.6859147 \\ 0.1552917 \\ 0.0615835 \\ 0.0297998 \\ 0.0159094 \\ 0.0089430 \\ 0.0051580 \\ 3.710e-02 \\ 5.254e-09 \\ 3.284e-15 \\ 2.323e-21 \\ 0.000300 \end{bmatrix}^T$$

Example 3.5. We use the presented methods to fit data from a theoretical Pareto I ($\alpha = 1.5, B = 4$) [74] distribution to APHDs. The results are shown in Fig. 3.7 where the corresponding PHD representations are given in Figs. 3.8 and 3.9. Using the D&C-EM method [146] the data trace with high coefficient of variation $C = 8.89$ (see Fig. 3.7) was separated in three partitions, in order to fit each partition into a hyper-exponential PHD using the EM algorithm [156]. We generate the final result by combining the weight $w_1 = 0.9626, w_2 = 0.0371, w_3 = 0.0003$ of each partition to the entire continuous data histogram with its corresponding fitted hyper-exponential PHD with parameters visualized in Fig. 3.8. The order of the resulting PHD is obtained as the sum of phases over all partitions, i.e. as the sum of 7, 4, and 1 phases.

a

$$\begin{array}{l}\lambda(1)=1.5226\\ \lambda(2)=1.6625\\ \lambda(3)=1.8018\\ \lambda(4)=2.1916\\ \lambda(5)=2.3485\\ \lambda(6)=2.7586\\ \lambda(7)=4.1700\\ \lambda(8)=5.0370\end{array}\qquad \pi^1=\begin{bmatrix}0.3553\\0.0425\\0.0460\\0.1079\\0.1451\\0.0961\\0.0860\\0.0772\end{bmatrix}^T \qquad \boldsymbol{D}_0^1=\begin{bmatrix}-\lambda(1) & \lambda(1) & 0 & \dots & 0\\ 0 & -\lambda(2) & \lambda(2) & & \\ & & \ddots & \ddots & \vdots\\ \vdots & & & -\lambda(7) & \lambda(7)\\ 0 & & & & -\lambda(8)\end{bmatrix}.$$

b

$$\mathbf{D}_0^2=diag\begin{bmatrix}-0.0557\\-0.0128\\-0.0029\\-0.0006\\-0.0001\\-3.5e-005\\-8.2e-006\\-1.8e-006\\-3.9e-007\\-6.2e-008\end{bmatrix}\quad \pi^2=\begin{bmatrix}0.0389\\0.0042\\0.0004\\5.2e-005\\5.7e-006\\6.3e-007\\6.9e-008\\7.6e-009\\8.1e-010\\7.2e-011\end{bmatrix}^T$$

c

$$\boldsymbol{D}_0=\begin{bmatrix}\mathbf{D}_0^1 & \mathbf{0}\\ \mathbf{0} & \mathbf{D}_0^2\end{bmatrix},\quad \pi=\left[\pi^1\,\pi^2\right].$$

Fig. 3.9 Applying the PHfit approach [74] to the Pareto I distribution with density visualized in Fig. 3.7 is approximated using eight phases for fitting the main part, and ten phases for fitting the tail which are shown in Figs. 3.9a and 3.9b. The final PHD (see Fig. 3.9c) is obtained by combining the series canonical form part and the hyper-exponential part of the PHD (**a**) The PHD in series canonical form approximating the main part of the distribution. (**b**) The hyper-exponential PHD approximating the tail obtained using the method [74] (**c**) The sub generator $\boldsymbol{D}_0$ of order 18 obtained as a combination of two parts of the PHD. The $\mathbf{D}_0^1$ is in series canonical form, in particular $0<\lambda(1)\leq\lambda(2)\leq\ldots\leq\lambda(7)\leq\lambda(8)$

3.2 Moments Based Fitting

Approaches for moment based fitting of PHDs can be further distinguished depending on whether they construct a PHD that exactly matches the given moments or whether they only try to approximate the moments. In the first case it is usually tried to directly relate the moments and the entries from $\boldsymbol{\pi}$ and $\mathbf{D}_0$ via some closed form equations. An important question in this regard is, of course, whether the moments can be reached by a PHD of a given order, i.e. whether the equations yield a valid PHD description. We will treat this issue and some approaches for moment matching in Sect. 3.2.1. In the second case it is tried to construct $\boldsymbol{\pi}$ and $\mathbf{D}_0$ using optimization techniques such that the difference between the moments of the PHD and the given moments from a trace is minimized. While these approaches only approximate the given moments, they always result in a valid PHD description. These techniques are considered in Sect. 3.2.2.

3.2.1 Closed Form Equations

Moment matching algorithms using closed form equations directly relate the entries of $(\boldsymbol{\pi}, \mathbf{D}_0)$ to (empirical) moments. This implies, that the order of the resulting PHD is determined by the number of moments used for the approach, i.e. it is known that a PHD of order n is completely determined by $2n-1$ moments [155]. As a consequence all higher order moments of a PHD can be computed in terms of the first $2n-1$ moments. Let $\mu_i, 1 \leq i \leq 2n-1$ be the moments of a $PHD(n)$. Then, the factorial moments are obtained by $r_i = \mu_i / i!$. Define the matrix

$$\mathbf{H}_{2n} = \begin{bmatrix} r_0 & r_1 & \cdots & r_n \\ r_1 & r_2 & \cdots & r_{n+1} \\ \vdots & \vdots & \ddots & \vdots \\ r_n & r_{n+1} & \cdots & r_{2n} \end{bmatrix}$$

Then, r_{2n} can be obtained by solving $det(\mathbf{H}_{2n}) = 0$, i.e. r_{2n} is a function of the $r_i, 0 \leq i \leq 2n-1$. It follows that

$$r_{2n} = -\frac{\sum_{i=1}^{n} r_{n+i-1} det_{i,n+1}(\mathbf{H}_{2n})}{det\left(\mathbf{H}'_{2n-2}\right)} \tag{3.32}$$

where $det_{i,j}(\mathbf{H}_{2n})$ is the (signed) subdeterminant of the element i, j and $\mathbf{H}'_{2n-2}$ is the $n \times n$ matrix that results from $\mathbf{H}_{2n}$ by deleting the last row and the last column. In a similar way one can obtain expressions for further moments, e.g.,

$$r_{2n+1} = -\frac{\sum_{i=1}^{n} r_{n+i} det_{i,n+1}(\mathbf{H}_{2n+1})}{det\left(\mathbf{H}'_{2n-1}\right)} \tag{3.33}$$

Unfortunately, explicit moment bounds are not known beyond $n = 2$, which makes it difficult to decide, whether a set of moments can be matched by a PHD. Telek and Heindl [154] give explicit moment bounds for APHDs of order 2. In fact, it is known that the classes of $APHD(2)$, $PHD(2)$ and even matrix exponential distributions of order 2 coincide [23, 80], such that these bounds are also valid for more general cases than APHDs. In particular, it can be shown that the moments of an $APHD(2)$ lie within the following range [154]:

$$\begin{array}{lll} \mu_1 : & 0 < \mu_1 < \infty & \\ C^2 : & 0.5 \leq C^2 < \infty & \\ \mu_3 : & 3\mu_1^3(3C^2 - 1 + \sqrt{2}(1 - C^2)^{3/2}) \leq \mu_3 \leq 6\mu_1^3 C^2 & \text{if } 0.5 \leq C^2 \leq 1 \\ & \frac{3}{2}\mu_1^3(1 + C^2)^2 < \mu_3 (< \infty) & \text{if } 1 < C^2 \end{array}$$

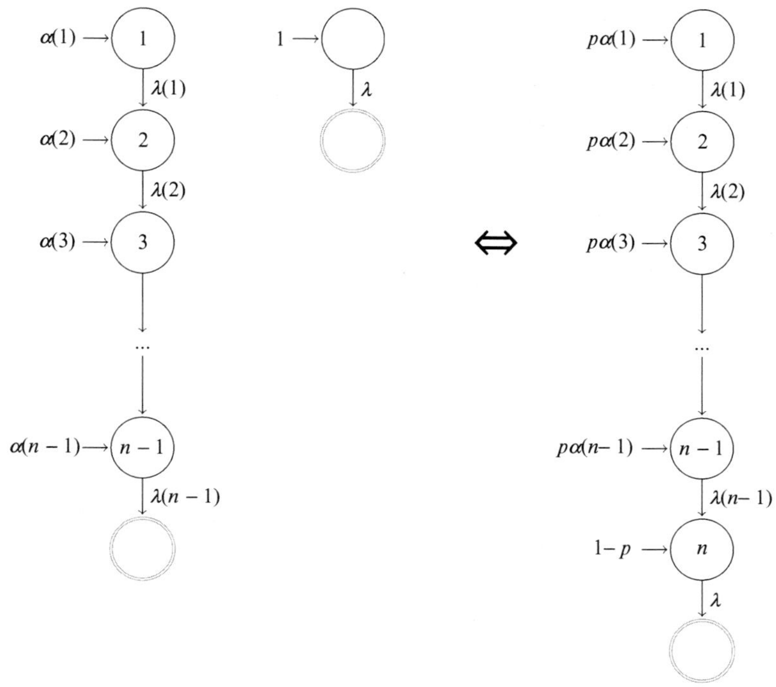

Fig. 3.10 Construction of an $APHD(n)$ from an $APHD(n-1)$ and an exponential distribution

Note, that instead of the second moment, the squared coefficient of variation C^2 is given, from which the second moment can be computed easily. Telek and Heindl [154] also developed expressions to match the parameters of an $APHD(2)$ according to these three moments. If the given moments are not feasible with an $APHD(2)$ the moment bounds enable us to adjust the moments to the closest value that lies within the bounds. Bobbio et al. [21] investigated the case to construct an APHD of minimal order that matches three given moments thereby extending work from [137, 138].

Horvàth and Telek [77] present an iterative approach for matching an arbitrary number of moments to an APHD. Although they do not give explicit bounds for the moments, the approach works in a way, that it only results in a valid APHD representation if the moments are feasible.

The approach constructs an APHD of order n in series canonical form and in each iteration the rate λ of one phase and the corresponding initial probability $1-p$ is computed resulting in an APHD of order $n-i$ to be constructed after the ith iteration of the algorithm. Figure 3.10 shows the key idea how an $APHD(n)$ is build from an $APHD(n-1)$ and an additional exponential phase. To simplify the matching problem and the obtained expressions the approach uses normalized moments which are defined as

$$n_i = \frac{\mu_i}{\mu_{i-1}\mu_1}, i \geq 1. \tag{3.34}$$

The (ordinary) moments μ_i can be computed by the normalized moments using

$$\mu_i = n_i \mu_{i-1}\mu_1 = \mu_1^i \prod_{k=1}^{i} n_k.$$

Note, that we only consider non-defective distributions with $\mu_0 = 1$, implying that $n_1 = 1$.

In particular, the approach from [77] starts with $2n-2$ normalized moments n_i, uses the two highest moments to compute λ and p and constructs an $APHD(n-1)$ with the remaining $2n-4$ normalized moments. The algorithm requires a relation between the (normalized) moments of the $APHD(n)$ and the $APHD(n-1)$.

Let μ_i be the moments of an $APHD(n)$ and μ_i' be the moments of an $APHD(n-1)$ related as shown in Fig. 3.10. Then, the moments μ_i, can be derived from the μ_i', the rate λ of the additional phase and the probability p [77], i.e.

$$\mu_i = i!\lambda^{-i}\left(1 + p\sum_{j=1}^{i}\frac{\lambda^j \mu_j'}{j!}\right).$$

The inverse relation is given by

$$\mu_i' = \frac{\lambda\mu_i - i\mu_{i-1}}{\lambda p}.$$

Similar expressions can be found for the relation of the normalized moments n_i of the $APHD(n)$ and n_i' of the $APHD(n-1)$ [77]:

$$n_i = \frac{i\left(1 + b\sum_{j=1}^{i}\frac{a^{j-1}}{j!}\prod_{k=1}^{j} n_k'\right)}{(1+b)\left(1 + b\sum_{j=1}^{i-1}\frac{a^{j-1}}{j!}\prod_{k=1}^{j} n_k'\right)} \tag{3.35}$$

where

$$a = \mu_1'\lambda \quad \text{and} \quad b = ap \tag{3.36}$$

are used to simplify the expression and eliminate the dependence on the first moment. Note, that a is the ratio of the means of the $APH(n-1)$ distribution and the exponential distribution that are combined into the $APH(n)$ distribution. The normalized moments n_i' of the $APHD(n-1)$ are a function of n_i, n_{i-1}, a and b:

$$n_i' = f(n_i, n_{i-1}, a, b) = \frac{n_{i-1}(1+b)(n_i(1+b)-i)}{a(n_{i-1}(1+b)-(i-1))}, \quad i > 1. \tag{3.37}$$

Algorithm 3 Iterative algorithm for moments matching

Input: moments $\mu_i, 1 \le i \le 2n-1$
Output: APHD of order n
1: compute normalized moments $n_i, 2 \le i \le 2n-1$ according to Eq. (3.34)
2: set $n_i^{(n)} = n_i, 2 \le i \le 2n-1$
3: **for** $j = n \rightarrow 2$ **do**
4: // construct equation for a using $n_{2j-2}^{(j)}$
5: take Eq. (3.35) for $n_{2j-2}^{(j)}$
6: substitute $n_{2j-2}^{(j-1)}$ by $\mathcal{F}_{n_{2j-2}^{(j-1)}}\left(n_2^{(j-1)}, \cdots, n_{2j-3}^{(j-1)}\right)$
7: substitute $n_i^{(j-1)}$ by $f(n_i^{(j)}, n_{i-1}^{(j)}, a, b)$ for $i = 1, \cdots, 2j-3$
8: // construct equation for b using $n_{2j-1}^{(j)}$
9: take Eq. (3.35) for $n_{2j-1}^{(j)}$
10: substitute $n_{2j-1}^{(j-1)}$ by $\mathcal{F}_{n_{2j-1}^{(j-1)}}\left(n_2^{(j-1)}, \cdots, n_{2j-2}^{(j-1)}\right)$
11: substitute $n_i^{(j-1)}$ by $f\left(n_i^{(j)}, n_{i-1}^{(j)}, a, b\right)$ for $i = 1, \cdots, 2j-2$
12: // compute solution
13: solve equations for a and $b \Rightarrow j$ solutions $\mathcal{S} = \{(a_1, b_1), \cdots (a_j, b_j)\}$
14: set $(a^{(j)}, b^{(j)}) = \max_{b_i}((a_i, b_i) \in \mathcal{S})$
15: // compute normalized moments of APH(j - 1) distribution according to Eq. (3.37)
16: $n_i^{(j-1)} = f(n_i^{(j)}, n_{i-1}^{(j)}, a^{(j)}, b^{(j)})$ for $i = 2, \cdots, 2j-3$
17: **end for**
18: set $\lambda(1) = 1, p(1) = 1$
19: compute $\lambda(i), p(i), i = 2, \cdots, n$ from $a^{(i)}, b^{(i)}$ according to Eq. (3.36)
20: scale $\lambda(i), p(i), i = 1, \cdots, n$ to match μ_1
21: **return** $\lambda(i), p(i), i = 1, \ldots, n$

Using these relations we can formulate the iterative Algorithm 3 for matching an arbitrary number of moments.

The algorithm starts with the $2n-1$ moments μ_i and iteratively determines the rates and probabilities of the n phases of an APHD in series canonical form. In the first step the normalized moments $n_i, 2 \le i \le 2n-1$ are computed according to Eq. (3.34) (recall that $n_1 = 1$). In each iteration j the algorithm maintains a set of normalized moments $n_i^{(j)}$ for the $APH(j)$ currently under construction. The moments $n_i^{(j)}, i = 1, \cdots, 2j-3$ will be used to compute the normalized moments for the next iteration in later steps of the algorithm. The two remaining normalized moments $n_{2j-2}^{(j)}$ and $n_{2j-1}^{(j)}$ are used to derive expressions for a and b. To determine a we use Eq. (3.35) for $n_{2j-2}^{(j)}$ and obtain an expression containing $n_i^{(j-1)}, 1 \le i \le 2j-2$. Since these are the normalized moments of an $APHD(j-1)$ (which is determined by $2j-3$ moments) the moment $n_{2j-2}^{(j-1)}$ is redundant and can be expressed in terms of the lower moments (line 6). Using $r_0 = 1$ and $r_i = 1/i! \prod_{j=1}^{i} n_j$ we can define a relation similar to Eq. (3.32) for the normalized moments, denoted by the function $\mathcal{F}_{n_{2j-2}^{(j-1)}}\left(n_2^{(j-1)}, \cdots, n_{2j-3}^{(j-1)}\right)$ in the algorithm. Finally, the unknown moments $n_i^{(j-1)}$ in the equation are substituted according to Eq. (3.37) resulting in an equation for a

containing b and the known $n_i^{(j)}$. The equation is linear in a [77] and thus, can be rearranged to get an explicit expression for a. In a similar way an expression for b can be obtained using the normalized moment $n_{2j-1}^{(j)}$ (lines 8–11). The equation for b can be arranged to a polynomial equation of order j [77] and therefore in iteration j we obtain j solutions for a, b. In fact, it is possible to arrange the equation for b in a way, such that it does not depend on a any longer [77]. However, depending on the order of the polynomial this might be difficult. Hence, we decided to solve a system with both equations and the two unknown variables a and b in Algorithm 3. It can be shown that a set of moments $n_i^{(j)}, 1 \leq i \leq 2j-1$ is feasible with an APHD of order j if and only if there exists a solution of the reduction step with moments $n_i^{(j-1)}, 1 \leq i \leq 2j-3$ and a, b such that a and b are real numbers ($0 < b \leq a$) and the normalized moments $n_i^{(j-1)}, 1 \leq i \leq 2j-3$ are feasible with an $APHD(j-1)$ [77]. The check for feasibility is omitted in Algorithm 3 and we implicitly assume that the moments are feasible. Moreover, it can be shown [77] that selecting the largest b (and the corresponding a and normalized moments) results in an APHD in series canonical form if the moments are feasible. If the moments are not feasible the procedure provides an improper APHD. Consequently, we select the solution with the largest b in line 14 as $(a^{(j)}, b^{(j)})$ and proceed with those values to compute the normalized moments for the next iteration $n_i^{(j-1)}, i = 1, \cdots, 2j-3$ according to Eq. (3.37) (line 16).

From the $(a^{(i)}, b^{(i)})$ pairs we can construct the probabilities and rates of the APHD. Since we have used one parameter less than required to determine the APHD (recall, the due to the normalized moments the mean was not used for fitting yet) we have a degree of freedom and can set the first rate to an arbitrary value (e.g. $\lambda(1) = 1$). Then, using Eq. (3.36) and $(a^{(j)}, b^{(j)})$ the parameters $\lambda(i)$ and $p(i)$ are added iteratively. In the final step of the algorithm (line 20) the rates are scaled to match the first moment and the rates and probabilities are returned.
To clarify the steps of the algorithm consider the following examples:

Example 3.6. Assume, that we want to construct an APHD of order 3 using the given moments

$$\mu_0 = 1, \mu_1 = 1.4567, \mu_2 = 4.7222, \mu_3 = 24.984, \mu_4 = 186.39, \mu_5 = 1796.1.$$

According to Eq. (3.34) this results in the normalized moments

$$n_2 = n_2^{(3)} = 2.2255, \quad n_3 = n_3^{(3)} = 3.6321,$$

$$n_4 = n_4^{(3)} = 5.1213, \quad n_5 = n_5^{(3)} = 6.6154,$$

which are computed in the first step of Algorithm 3. By definition $n_1 = 1$.
In the first iteration of the main loop (for $j = 3$) the parameters $a^{(3)}, b^{(3)}$, which determine the rate and the probability of the last phase $\lambda(3)$ and $p(3)$ and the normalized moments $n_2^{(2)}, n_3^{(2)}$ for the next iteration, have to be computed.

We start with the computation of $a^{(3)}$ and $b^{(3)}$. To construct the equation for $a^{(3)}$ we take Eq. (3.35) for the known value $n_4^{(3)} = 5.1213$, i.e.

$$n_4^{(3)} = \frac{4\left(1+b\left(1+\frac{1}{2}an_2^{(2)}+\frac{1}{6}a^2n_2^{(2)}n_3^{(2)}+\frac{1}{24}a^3n_2^{(2)}n_3^{(2)}n_4^{(2)}\right)\right)}{(1+b)\left(1+b\left(1+\frac{1}{2}an_2^{(2)}+\frac{1}{6}a^2n_2^{(2)}n_3^{(2)}\right)\right)}. \tag{3.38}$$

In Eq. (3.38) the unknown moment $n_4^{(2)}$ is substituted by $\mathcal{F}_{n_4^{(2)}}\left(n_2^{(2)},\cdots,n_3^{(2)}\right)$. This is done using the reduced moments r_i. According to Eq. (3.32) we can compute r_4 by

$$r_4 = -\frac{\sum_{i=1}^{2} r_{2+i-1}det_{i,3}(\mathbf{H}_4)}{det(\mathbf{H}_2')}$$

where

$$\mathbf{H}_4 = \begin{bmatrix} r_0 & r_1 & r_2 \\ r_1 & r_2 & r_3 \\ r_2 & r_3 & r_4 \end{bmatrix} \quad \text{and} \quad \mathbf{H}_2' = \begin{bmatrix} r_0 & r_1 \\ r_1 & r_2 \end{bmatrix},$$

i.e. we obtain

$$r_4 = -\frac{r_2\cdot(-1)^4\cdot(r_1r_3-r_2r_2)+r_3\cdot(-1)^5\cdot(r_0r_3-r_1r_2)}{r_0r_2-r_1r_1}.$$

Substituting $r_i = 1/i!\prod_{j=1}^{i} n_j^{(2)}$ we obtain after some simplifications

$$n_4^{(2)} = \frac{2n_2^{(2)}\left(9n_2^{(2)}+2\left(n_3^{(2)}-6\right)n_3^{(2)}\right)}{3\left(n_3^{(2)}-2\right)n_2^{(2)}}. \tag{3.39}$$

Then, $n_4^{(2)}$ in Eq. (3.38) is substituted according to Eq. (3.39) and the remaining $n_i^{(2)}$ are substituted using Eq. (3.37) resulting in a lengthy equation containing a and b as unknowns.

In a similar way we can construct the equation for $b^{(3)}$, i.e. we take Eq. (3.35) for the known value $n_5^{(3)} = 6.6154$:

$$n_5^{(3)} = \frac{5\left(1+b\left(1+\frac{1}{2}an_2^{(2)}+\frac{1}{6}a^2n_2^{(2)}n_3^{(2)}+\frac{1}{24}a^3n_2^{(2)}n_3^{(2)}n_4^{(2)}+\frac{1}{120}a^4n_2^{(2)}n_3^{(2)}n_4^{(2)}n_5^{(2)}\right)\right)}{(1+b)\left(1+b\left(1+\frac{1}{2}an_2^{(2)}+\frac{1}{6}a^2n_2^{(2)}n_3^{(2)}+\frac{1}{24}a^3n_2^{(2)}n_3^{(2)}n_4^{(2)}\right)\right)}. \tag{3.40}$$

In Eq. (3.40) the unknown moment $n_5^{(2)}$ has to be substituted by $\mathcal{F}_{n_5^{(2)}}(n_2^{(2)},\cdots,n_4^{(2)})$ using the reduced moments and following the procedure as described for the computation of $a^{(3)}$. In the expression for $n_5^{(2)}$ we can additionally replace $n_4^{(2)}$ by the term from Eq. (3.39) resulting in

$$n_5^{(2)} = \frac{45n_2^{(2)}n_2^{(2)}(2n_3^{(2)}-3)+10n_2^{(2)}n_3^{(2)}((n_3^{(2)}-12)n_3^{(2)}+9)+60n_3^{(2)}n_3^{(2)}}{3(n_2^{(2)}-2)(9n_2^{(2)}+2(n_3^{(2)}-6)n_3^{(2)})}. \quad (3.41)$$

As before, we substitute $n_5^{(2)}$ in Eq. (3.40) according to Eq. (3.41) and the remaining $n_i^{(2)}$ according to Eq. (3.37) resulting in a second equation containing a and b as unknowns.

Solving the two equations we obtain three solutions

$$\mathcal{S} = \{(a_1,b_1),\cdots(a_j,b_j)\} = \{(0.512,-0.272),(1.945,0.465),(2.202,0.931)\}.$$

According to the algorithm we choose the tuple with the largest b_i and set $a^{(3)} = 2.202, b^{(3)} = 0.931$. Since $0 < b^{(3)} \le a^{(3)}$ the solution is feasible.

Now, that we have solutions for a and b we can use Eq. (3.37) to compute the normalized moments for the next iteration. In particular we obtain

$$n_2^{(2)} = \frac{n_1^{(3)}(1+b^{(3)})\left(n_2^{(3)}(1+b^{(3)})-2\right)}{a^{(3)}\left(n_1^{(3)}(1+b^{(3)})-1\right)} = 2.164$$

$$n_3^{(2)} = \frac{n_2^{(3)}(1+b^{(3)})\left(n_3^{(3)}(1+b^{(3)})-3\right)}{a^{(3)}\left(n_3^{(3)}(1+b^{(3)})-2\right)} = 3.409$$

In the next (and last) iteration for $j=2$ we use $n_2^{(2)}$ and $n_3^{(2)}$ to compute values for $a^{(2)}$ and $b^{(2)}$, respectively. Applying the same steps as before, we finally obtain the two equations

$$2.164 = \frac{2((a+1)b+1)}{(b+1)^2}$$

$$3.409 = \frac{3\left(b\left(\frac{a(b+1)(2.16356(b+1)-2)}{2b}+\frac{(b+1)(2.16356(b+1)-2)}{2b}+1\right)+1\right)}{(b+1)\left(b\left(\frac{(b+1)(2.16356(b+1)-2)}{2b}+1\right)+1\right)}$$

which we solve for a and b resulting in two solutions

$$\mathcal{S} = \{(a_1,b_1),(a_2,b_2)\} = \{(0.497,-0.169),(2.011,0.671)\}.$$

Consequently, we set $a^{(2)} = 2.011, b^{(2)} = 0.671$. Again, the solution is feasible. The computation of normalized moments for the next iteration (line 16 in Algorithm 3 is skipped (in fact, there are no remaining normalized moments except for $n_1 = 1$).

Finally, we have to construct the APHD from the parameters $a^{(i)}, b^{(i)}$. The construction is done stepwise, i.e. starting with an $APHD(1)$ with a single rate we add one additional rate at a time according to $a^{(i)}, b^{(i)}$ until we have reached an

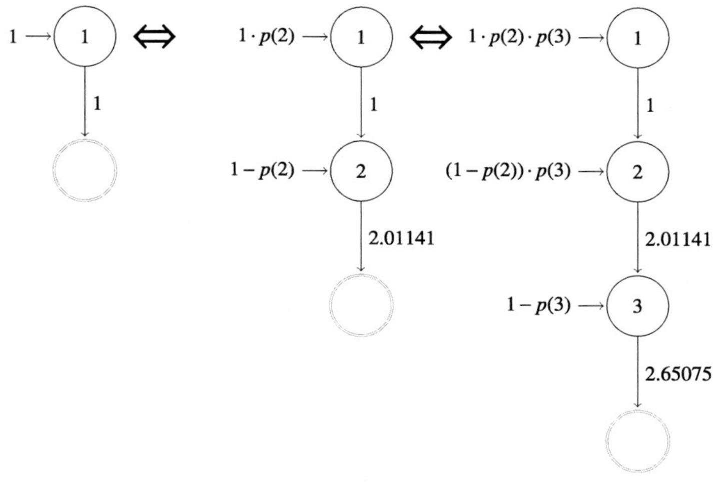

Fig. 3.11 Construction of the $APH(n)$ from Example 3.6

$APHD(n)$. Recall, that we have used $2n-2$ normalized moments for fitting and the empirical mean value has not been used yet. Consequently, we have a degree of freedom for the parameters, implying that we can choose an arbitrary rate for $\lambda(1)$. The rate is later adjusted when scaling the $\lambda(i)$ according to the mean. The steps for constructing the distribution are shown in Fig. 3.11. We start with a single phase with rate $\lambda(1) = 1$ and initial probability 1. Using $a^{(2)} = 2.011$ and $b^{(2)} = 0.671$ the second phase is constructed according to Eq. (3.36). From Eq. (3.36) we have that

$$p(2) = b^{(2)}/a^{(2)} = 0.33366.$$

$p(2)$ is the probability of the $APHD(1)$ and $1-p(2)$ the probability of the additional phase. The rate $\lambda(2)$ is computed from $\lambda(2) = a^{(2)}/\mu_1' = 2.01141$ where $\mu_1' = 1$ is the mean of the $APHD(1)$. Using $a^{(3)} = 2.202, b^{(3)} = 0.931$ we can add the third phase. We obtain $p(3) = 0.423$ which is the probability of the $APHD(2)$ constructed so far. Hence, the initial probabilities of the first two phases are multiplied with $p(3)$ and the third phase has the initial probability $1-p(3)$. It is easy to verify (cf. Eq. (2.14)) that the $APHD(2)$ has mean $\mu_1' = 0.8308$ and therefore we get $\lambda(3) = 2.65075$ resulting in the $APHD(3)$ shown in the left of Fig. 3.11. According to Eq. (2.14) we can compute the mean of the distribution which is $\mu_1^* = 0.728$. The desired mean is $\mu_1 = 1.4567$ and therefore we have to scale the rates $\lambda(i)$. Setting $\lambda'(i) = \lambda(i)\mu_1^*/\mu_1$ we obtain the distribution with the desired moments.

Example 3.7. The second example treats the case where the moments are not feasible for an $APHD(3)$. Let

$$\mu_1 = 0.42, \mu_2 = 0.504, \mu_3 = 1.2552, \mu_4 = 4.80768, \mu_5 = 23.9126.$$

The moments are generated from a matrix exponential distribution for which it is known that it cannot be represented by a PH distribution of order 3. The normalized moments are

$$n_1 = 1, n_2 = 2.85714, n_3 = 5.92971, n_4 = 9.11955, n_5 = 11.8425.$$

Following the steps from Algorithm 3 we obtain after the first iteration $a^{(3)} = 0.322222, b^{(3)} = -0.58$ which is not a valid solution because $0 < b^{(3)} \leq a^{(3)}$ does not hold. Consequently the algorithm would stop because no $APHD(3)$ can be constructed with the desired moments.

An advantage of the presented method is that one can easily check during the construction if the moments are feasible. An older approach that lacks this nice property was proposed in [110]. The algorithm constructs a matrix $\mathbf{K}$ from the factorial moments $(r_0, r_1, \cdots r_{2n-1})$. $\mathbf{K}$ is not a valid subgenerator of a PHD. However, from the vector-matrix pair $(\mathbf{e}_1, \mathbf{K})$, where $\mathbf{e}_1 = [1, 0, \cdots, 0]$, the moments can be computed as $\mu_i / i! = \mathbf{e}_1 \mathbf{K}^i \mathbf{e}_1^T$. To obtain a closing vector $\mathbb{1}$ one can multiply with a matrix $\mathbf{T}$ with $\mathbf{T}(i,j) = 1$ if $i \geq j$ and 0 otherwise resulting in $\mu_i / i! = \mathbf{e}_1 \mathbf{K}^i \mathbf{e}_1^T = \mathbf{e}_1 (\mathbf{TKT}^{-1})^n \mathbb{1}$ which corresponds to Eq. (2.14) for computing the moments of a PHD. Using the similarity transformations presented in Sect. 2.2.1 one can try to transform this representation into a valid representation for a PHD (and thereby check if the moments are feasible). The computation of an adequate transformation matrix $\mathbf{B}$ has to be done with a non-linear optimization algorithm. If the optimization fails, i.e. no valid representation can be found, it is open, whether the optimization algorithm failed or no valid solution exists.

3.2.2 *Least Squares Based Techniques*

In many cases an exact matching of moments by a PHD as presented in the previous section is not possible, because the set of moments to be matched cannot be modeled by a PHD of the corresponding order. In these cases an approximate fitting of the moments using least squares based techniques is possible. The resulting PHD does not exactly match the given moments, but the approaches always result in a valid PHD description. Experiments show, that the approximation is sufficiently close in most cases and in fact, the moments are usually estimated from a trace and as estimates they should be interpreted in the context of a confidence interval, which justifies an approximate fitting.

Let $\mathcal{M}$ be a set of moments to be approximated and $\hat{\mu}_i, i \in \mathcal{M}$ be the moments estimated from the trace. Then we have to solve the following optimization problem

$$\min_{\boldsymbol{\pi}, \mathbf{D}_0} \left(\sum_{i \in \mathcal{M}} \left(\beta_i \frac{\mu_i}{\hat{\mu}_i} - \beta_i \right)^2 \right) \tag{3.42}$$

where $\mu_i, i \in \mathcal{M}$ are the moments of the valid PHD $(\boldsymbol{\pi}, \mathbf{D}_0)$ and the β_i are optional weights, which can for example be used to privilege lower order moments. Note, that in contrast to the approaches from Sect. 3.2.1 there is no relation between the order of the PHD and the number of moments to be considered.

Buchholz and Kriege [32] present an approach that solves Eq. (3.42) for APHDs in series canonical form. The approach has to optimize $2n-1$ parameters for an APHD of order n, i.e. $n-1$ parameters in the vector $\boldsymbol{\pi}$ and n parameters in matrix $\mathbf{D}_0$ (cf. Fig. 2.14).

To reduce the effort for optimization the approach from [32] divides the minimization problem from Eq. (3.42) into two smaller minimization problems, which are iteratively solved. This can be interpreted as an alternating least squares approach (ALS) [100].

First, assume that $\mathbf{D}_0$ is given and we want to optimize for $\boldsymbol{\pi}$. According to Eq. (2.14) the moments of a PHD are computed as $\mu_i = i!\boldsymbol{\pi}(-\mathbf{D}_0)^{-i}\mathbb{1}$. Since $\mathbf{D}_0$ is known, the ith conditional moments $m_i = i!(-\mathbf{D}_0)^{-i}\mathbb{1}$ are known as well. Since $\mu_i = \boldsymbol{\pi} m_i$, Eq. (3.42) becomes

$$\min_{\boldsymbol{\pi}:\boldsymbol{\pi}=1,\boldsymbol{\pi}\geq 0}\left(\sum_{i\in\mathcal{M}}\left(\beta_i\frac{\boldsymbol{\pi} m_i}{\hat{\mu}_i}-\beta_i\right)^2\right) \tag{3.43}$$

which is a non-negative least squares problem with a single linear constraint. For the solution of non-negative least squares problems efficient algorithms exist [107].

Now assume, that $\boldsymbol{\pi}$ is given and we are looking for the optimal $\mathbf{D}_0$. Since the approach constructs an APHD in series canonical form, the unknown elements in $\mathbf{D}_0$ are the diagonal entries $\lambda(r), r = 1, \cdots, n$ (cf. Fig. 2.14). This optimization step is more difficult, because for moment μ_i the matrix $(-\mathbf{D}_0)^{-i}$ is required. Therefore [32] suggests to optimize for a single rate $\lambda(r)$ and keep the other rates fixed. Assume that $\lambda(r)$ is modified by a factor Δ such that $\lambda(r)$ becomes $\lambda(r)/(1+\lambda(r)\Delta)$. The moment matrix $\mathbf{M} = -\mathbf{D}_0^{-1}$ becomes $\mathbf{M}_{\Delta,r} = \mathbf{M} + \Delta\mathbf{E}_r$ where $\mathbf{E}_r$ is a $n \times n$ matrix with 1 in position $(1,r), \cdots, (r,r)$ and 0 elsewhere. Then, the moments are given by $\mu_i(\Delta, r) = \boldsymbol{\pi}(\mathbf{M} + \Delta\mathbf{E}_r)^i\mathbb{1}$ and the minimization problem for rate $\lambda(r)$, fixed rates $\lambda(s), s \neq r$ and fixed $\boldsymbol{\pi}$ becomes

$$\min_{\Delta}\left(\sum_{i\in\mathcal{M}}\left(\beta_i\frac{\mu_i(\Delta,r)}{\hat{\mu}_i}-\beta_i\right)^2\right). \tag{3.44}$$

Equation 3.44 is not a non-negative least squares problem and has to be solved with standard optimization techniques.

The complete approach iterates between optimization of $\boldsymbol{\pi}$ as in Eq. (3.43) and optimization of $\lambda(r), r = 1, \cdots, n$ as in Eq. (3.44) until convergence is reached.

Example 3.8. Consider again Example 3.7. It was shown that no exact matching of the moments is possible with an *APHD*(3). However, using the least squares based technique described above we can fit an APHD that approximates the moments. After 39 iterations the approach converges to an *APHD*(3) with

$$\pi = \begin{bmatrix}0.125 & 0.517 & 0.358\end{bmatrix}, \qquad \mathbf{D}_0 = \begin{bmatrix} -0.9908 & 0.9908 & 0 \\ 0 & -4.333 & 4.333 \\ 0 & 0 & 6.874 \end{bmatrix}.$$

The corresponding moments are

$$\mu_1 = 0.42, \mu_2 = 0.50388, \mu_3 = 1.2561, \mu_4 = 4.8038, \mu_5 = 23.9197,$$

which are very close to the original moments.

3.3 Concluding Remarks

Although PHDs are known for a long time, computational algorithms to match the parameters according to measured data have been mainly developed during the last two decades. Early work [148] uses only two or at most three moments which are matched by a PHD. If higher order moments are used for parameter fitting, then the resulting equations become non-linear and hard to solve. By restricting the class of PHDs to APHDs or compositionally generated PHDs, matching of a small number of moments is nowadays possible. However, the estimation of higher order moments from a trace data is often unreliable since the estimators are sensitive for outliers. This aspect has not been investigated in the context of moment based fitting.

An alternative to moment based approaches is the approximation of the empirical density function by the PHD. Different fitting algorithms for this purpose have been proposed [6, 18, 35, 58, 87, 89, 94, 134, 135, 139, 156] but it seems that EM algorithms are most appropriate and it seems also that the class of PHDs should be restricted to APHDs or even more restricted subclasses for an efficient fitting approach. From a practical point of view, the restriction to APHDs seems not even to reduce the fitting quality, measured in the value of the likelihood function, for a fixed number of phases. Recent results indicate that for more complex and multimodal empirical densities modern EM algorithms can be applied to fit an APHD in a good quality and with an acceptable effort. The approach may sometimes even be used for heavy tailed data sets [74, 76].

Much less attention has been paid to fitting discrete PHDs in the literature. However, the interested reader is referred to [20] that presents an ML estimation procedure for discrete PHDs. The fitting of more general ME distributions is a more or less open problem which seems to be much more complicated than parameter fitting for PHDs. Some first approaches can be found in [54].

Chapter 4
Markovian Arrival Processes

PHDs can be extended to describe correlated inter-event times. The resulting models are denoted as *Markovian Arrival Processes* (MAPs) and have been introduced in the pioneering work of Neuts [124]. MAPs are a very flexible and general class of stochastic processes. In this chapter we first introduce the general model and its analysis, then the specific case of MAPs with only two states is considered because it allows one to derive some analytical results and canonical representations. The last section extends the model class to stochastic processes generating different event types.

4.1 Definition and Basic Results

First, the class of MAPs is defined, then basic quantities characterizing the generated event stream are analyzed, equivalent representations are considered and finally, a different viewpoint is taken by interpreting the MAP as a counting process over some finite interval.

4.1.1 Definition of MAPs

Formally, a MAP can be interpreted as an irreducible Markov chain where some transitions are marked. Marked transitions describe events.

Definition 4.1. A Markovian Arrival Process (MAP) $(\boldsymbol{\pi}, \mathbf{D}_0, \mathbf{D}_1)$ is an irreducible Markov chain with a finite state space $\mathcal{S}$, initial vector $\boldsymbol{\pi}$ and generator matrix $\mathbf{Q}$ which can be represented as $\mathbf{Q} = \mathbf{D}_0 + \mathbf{D}_1$ where $\mathbf{D}_1 \geq \mathbf{0}$, $\mathbf{D}_1 \neq \mathbf{0}$, $\mathbf{D}_0(i,j) \geq 0$ for $i \neq j$ and $(\boldsymbol{\pi}, \mathbf{D}_0)$ is a valid PHD.

P. Buchholz et al., *Input Modeling with Phase-Type Distributions and Markov Models: Theory and Applications*, SpringerBriefs in Mathematics,
DOI 10.1007/978-3-319-06674-5_4,

a

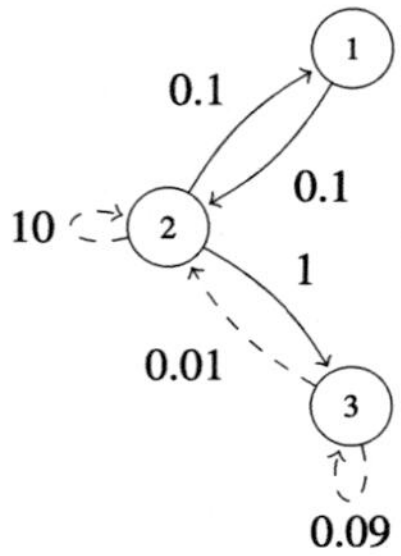

b

$$\mathbf{D}_0 = \begin{bmatrix} -0.1 & 0.1 & 0 \\ 0.1 & -11.1 & 1 \\ 0 & 0 & -0.1 \end{bmatrix}, \mathbf{D}_1 = \begin{bmatrix} 0 & 0 & 0 \\ 0 & 10 & 0 \\ 0 & 0.01 & 0.09 \end{bmatrix}$$

Fig. 4.1 A 3-state MAP. The *dashed transition arrows* correspond to state transitions generating an event (**a**) The state transition diagram for the MAP (**b**) The matrices for the MAP

$\mathcal{S}$ is the state space of the MAP and $n = |\mathcal{S}|$ the size of the state space or the order of the MAP. The interpretation of the behavior of a MAP is as follows. The process starts with probability $\boldsymbol{\pi}(i)$ in state i, resides there an exponentially distributed time with rate $\lambda(i) = \sum_{j \neq i} \mathbf{D}_0(i,j) + \sum_j \mathbf{D}_1(i,j)$, generates then an event with probability $\sum_j \mathbf{D}_1(i,j)/\lambda(i)$ and chooses state j as successor state with probability $(\mathbf{D}_0(i,j) + \mathbf{D}_1(i,j))\,/\,\lambda(i)$ if $i \neq j$ and with probability $\mathbf{D}_1(i,i)/\lambda(i)$ if $i = j$. Often the initial vector is not part of the MAP definition. In this case $\mathbf{D}_0 + \mathbf{D}_1$ is assumed to be an irreducible generator matrix such that $\mathbf{P}_s = (-\mathbf{D}_0)^{-1}\mathbf{D}_1$ is an irreducible stochastic matrix with unique left eigenvector $\boldsymbol{\pi}_s \mathbf{P}_s = \boldsymbol{\pi}_s$ and $\boldsymbol{\pi}_s \mathbb{1} = 1$. Vector $\boldsymbol{\pi}_s$ includes the stationary distribution after an event and is then used as initial vector of the MAP. In this case we denote a MAP as $(\mathbf{D}_0, \mathbf{D}_1)$ with the implicit definition of the initial vector $\boldsymbol{\pi}_s$.

Example 4.1. Figure 4.1 shows a simple example of a MAP with 3 states. A MAP may contain transitions that start and end in the same state, in the example the transitions starting and ending in the states 2 and 3, but such transitions have to generate events. The initial distribution for the example MAP has not been defined yet. If we choose $\boldsymbol{\pi} = (1,0,0)$, then the behavior of the MAP is as follows. It starts in state 1 and resides there for an exponentially distributed time with a mean duration of 10. Afterwards, the state changes to state 2 without generating an event. In state 2 the MAP generates events with rate 10. Concurrently an exponential distribution with rate 1.1 is running. If this distribution elapses, the state changes with probability $0.1/1.1$ back to state 1 and with probability $1/1.1$ state 3 is entered. In state 3 the process generates events with an exponentially distributed inter-event time with rate 0.1. After generating an event the process stays with probability 0.9 in state 3 and moves with probability 0.1 simultaneously to state 2. It is easy to see that the inter-event times of this MAP are positively correlated because in state 2 events with a high rate are generated whereas the event generation rate is much lower in state 3. Measures for the correlation will be analyzed below.

Even if the definition of a MAP differs from the definition of a PHD, because the former uses irreducible CTMCs and the latter absorbing CTMCs, every PHD can be

a

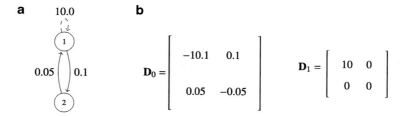

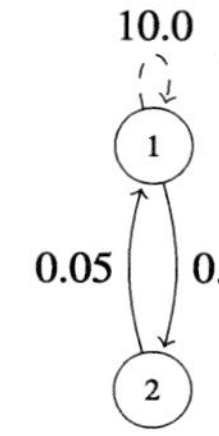

b

$$\mathbf{D}_0 = \begin{bmatrix} -10.1 & 0.1 \\ 0.05 & -0.05 \end{bmatrix} \qquad \mathbf{D}_1 = \begin{bmatrix} 10 & 0 \\ 0 & 0 \end{bmatrix}$$

Fig. 4.2 An IPP with 2 states (**a**) The state transition diagram for the IPP (**b**) The matrices for the IPP

represented by a MAP. Let $(\boldsymbol{\pi}, \mathbf{D}_0)$ be a PHD, then $(\boldsymbol{\pi}, \mathbf{D}_0, \mathbf{d}_1\boldsymbol{\pi})$ (where $\mathbf{d}_1 = -\mathbf{D}_0 \mathbb{1}$) is a MAP with the same behavior. That is, events are generated with independently and identically distributed inter-event times and the distribution equals PHD $(\boldsymbol{\pi}, \mathbf{D}_0)$.

There are several subclasses of MAPs that put some restrictions on the matrices $\mathbf{D}_0$ and $\mathbf{D}_1$. If $\mathbf{D}_0$ can be reordered to an upper triangular matrix by symmetric row and column permutations, then the inter-event times of the MAP are given by an APHD. Among the class of MAPs with acyclic matrix $\mathbf{D}_0$ one can define further subclasses by considering for example MAPs with matrices $\mathbf{D}_0$ describing hyper-Erlang or Coxian-distributions. If matrix $\mathbf{D}_1$ is a diagonal matrix, then the process is denoted as a *Markov Modulated Poisson Process* (MMPP) [59] because matrix $\mathbf{D}_1$ describes up to n Poisson processes for an MMPP with n states that are selected by a background Markov process defined by $\mathbf{D}_0$. A specific case of an MMPP is an *Interrupted Poisson Process* (IPP) [101] where diagonal elements of the diagonal matrix $\mathbf{D}_1$ are either 0 or λ, the rate of the basic Poisson process. On and off times of the Poisson process are given by a PHD.

Example 4.2. Figure 4.2 shows a typical IPP with two states, one *on* state (state 1) where events are generated and one *off* state (state 2) without event generation. The behavior of the process is as follows. After entering the *on* state, a geometrically distributed number of events is generated before the *off* state is entered. The mean number of events generated between entering and leaving the *on* state is $10/0.1 = 100$ and the mean inter-event time equals 0.1. The sojourn time in the *on* state is exponentially distributed with mean 10 and the sojourn time in the *off* state is exponentially distributed with mean 20.

IPPs are often used to model packetized voice traffic in computer networks [83]. It is easy to extend the basic two state model by modeling the times of the *on* and *off* phase by PHDs $(\boldsymbol{\pi}^{on}, \mathbf{D}_0^{on})$ and $(\boldsymbol{\pi}^{off}, \mathbf{D}_0^{off})$. If λ is the event generation rate in the *on* state, then the resulting extended IPP is described by the matrices

$$\mathbf{D}_0 = \begin{bmatrix} \mathbf{D}_0^{on} - \lambda\mathbf{I} & -\mathbf{D}_0^{on}\mathbb{1}\boldsymbol{\pi}^{off} \\ -\mathbf{D}_0^{off}\mathbb{1}\boldsymbol{\pi}^{on} & \mathbf{D}_0^{off} \end{bmatrix} \text{ and } \mathbf{D}_1 = \begin{bmatrix} \lambda\mathbf{I} & \mathbf{0} \\ \mathbf{0} & \mathbf{0} \end{bmatrix}$$

and initial vector $[\boldsymbol{\pi}^{on}, \mathbf{0}]$ or $[\mathbf{0}, \boldsymbol{\pi}^{off}]$, depending whether the process starts in the *on* or *off* state.

Observe that the inter-event times of an IPP are uncorrelated, as long as the event rate in the *off* states is zero and all *on* states have the same rate. However, if different inter generation rates are introduced, the IPP becomes an MMPP that describes correlated arrivals. For example, by assuming event rate 1 in the second state, we obtain an MMPP with the matrices

$$\mathbf{D}_0 = \begin{bmatrix} -10.1 & 0.1 \\ 0.05 & -1.05 \end{bmatrix} \text{ and } \mathbf{D}_1 = \begin{bmatrix} 10 & 0 \\ 0 & 1 \end{bmatrix}.$$

The embedded stationary vector of the MMPP equals $\boldsymbol{\pi}_s = \left[\frac{5}{6}, \frac{1}{6}\right]$ and equals the initial vector $\boldsymbol{\pi}$. The moments of the inter-event time are $E[T] = 0.25$, $E[T^2] = 0.33726$ and the first three coefficients of autocorrelation are $\rho_1 = 0.3644$, $\rho_2 = 0.34378$ and $\rho_3 = 0.32432$.

4.1.2 Analysis of MAPs

For the analysis of a MAP $(\boldsymbol{\pi}, \mathbf{D}_0, \mathbf{D}_1)$ we assume that $\boldsymbol{\pi} = \boldsymbol{\pi}_s$ and $\boldsymbol{\pi}_s$ is the unique solution of $\boldsymbol{\pi}_s \mathbf{P}_s = \boldsymbol{\pi}_s$ subject to $\boldsymbol{\pi}_s \mathbb{1} = 1$. Thus, the stationary inter-event time of a MAP is distributed according to PHD $(\boldsymbol{\pi}_s, \mathbf{D}_0)$ and can be analyzed with the equations presented in Sect. 2.1.3. If $\boldsymbol{\pi} \neq \boldsymbol{\pi}_s$, the MAP has a transient initial phase where the distribution differs from the stationary inter-event distribution. The inter-event times of a MAP are usually dependent.

In the sequel we assume $\boldsymbol{\pi}_s = \boldsymbol{\pi}$ to avoid the distinction between initial and stationary phase. However, the following analysis steps can be applied to any of the two phases, if the vectors differ. Since the distribution between event times can be analyzed with methods for PHDs, only measures that describe the dependencies between events are introduced. Let $X_1, \ldots, X_k$ be a sequence of k consecutive inter-event times and X an arbitrary inter-event time.

The joint density of a MAP generating k consecutive events with inter-event times x_i is given by

$$f(x_1, x_2, \ldots, x_k) = \boldsymbol{\pi} e^{\mathbf{D}_0 x_1} \mathbf{D}_1 e^{\mathbf{D}_0 x_2} \mathbf{D}_1 \ldots e^{\mathbf{D}_0 x_k} \mathbf{D}_1 \mathbb{1} \text{ for } x_1, \ldots, x_k \geq 0. \tag{4.1}$$

Equation 4.1 can be evaluated using uniformization [151]. Let $\alpha \geq \max_i(|\mathbf{D}_0(i,i)|)$, $\mathbf{P}_0 = \mathbf{D}_0/\alpha + \mathbf{I}$ and $\mathbf{P}_1 = \mathbf{D}_1/\alpha$, then

$$f(x_1, x_2, \ldots, x_k) = \boldsymbol{\pi} \left(\prod_{i=1}^{k} \left(\sum_{l=0}^{\infty} \beta(\alpha x_i, l) \mathbf{P}_0^l \right) \mathbf{P}_1 \right) \mathbb{1} \tag{4.2}$$

where $\beta(q, l)$ is the probability of l events of a Poisson process with parameter q.

The joint moments of k consecutive events and orders i_l $(1 \leq l \leq k)$ are defined as

$$E[X_1^{i_1}, X_2^{i_2}, \ldots, X_k^{i_k}] = \int_0^\infty \ldots \int_0^\infty (x_1)^{i_1} \ldots (x_k)^{i_k} f(x_1, x_2, \ldots, x_k) dx_1 \ldots dx_k \tag{4.3}$$

and can be computed for a MAP by

$$E[X_1^{i_1}, X_2^{i_2}, \ldots, X_k^{i_k}] = i_1! i_2! \ldots i_k! \pi(-\mathbf{D}_0)^{-i_1} \mathbf{P}_s (-\mathbf{D}_0)^{-i_2} \ldots \mathbf{P}_s (-\mathbf{D}_0)^{-i_k} \mathbb{1}. \quad (4.4)$$

For joint moments of two consecutive inter-event times, which will later be used in fitting approaches, we use the notation $\mu_{kl} = E[X_1^k, X_2^l]$. Often one is interested in first order properties between the first and kth event which are subsumed in the coefficient of autocorrelation of lag k defined as

$$\rho_k = \frac{E[X_1, X_{1+k}] - (E[X])^2}{E[X^2] - (E[X])^2} = \frac{\pi(-\mathbf{D}_0)^{-1} \mathbf{P}_s^k (-\mathbf{D}_0)^{-1} \mathbb{1} - \left(\pi(-\mathbf{D}_0)^{-1} \mathbb{1}\right)^2}{2\pi(-\mathbf{D}_0)^{-2} \mathbb{1} - \left(\pi(-\mathbf{D}_0)^{-1} \mathbb{1}\right)^2}. \quad (4.5)$$

ρ_k lies between -1 and 1. For uncorrelated event times the coefficients of autocorrelation are all 0, positive values indicate a positive correlation between event times and negative values a negative correlation. For a MAP $\lim_{k\to\infty} \rho_k = 0$ holds.

Example 4.3. For the example shown in Fig. 4.1 matrix $\mathbf{P}_s$ equals

$$\begin{bmatrix} 0.0 & 0.91818 & 0.08182 \\ 0.0 & 0.91818 & 0.08182 \\ 0.0 & 0.1 & 0.9 \end{bmatrix}$$

such that $\pi_s = [0, 0.55, 0.45]$. We assume that $\pi = \pi_s$ and obtain $E[X] = 5.1$ and $C^2 = 2.8915$ for the inter-event time distributions. The first three autocorrelation coefficients are 0.26217, 0.2145 and 0.1755.

4.1.3 Equivalent Representations of MAPs

Equivalence of PHDs has been introduced in Sect. 2.2. The corresponding equivalence relations can all be extended to MAPs. We begin with similarity transformations for some MAP $(\pi, \mathbf{D}_0, \mathbf{D}_1)$ of order n. Let $\mathbf{B}$ be a non-singular $n \times n$ matrix with unit row sums, then $(\pi\mathbf{B}, \mathbf{B}^{-1}\mathbf{D}_0\mathbf{B}, \mathbf{B}^{-1}\mathbf{D}_1\mathbf{B})$ is an equivalent representation of the same MAP. Equivalence can be easily proved by substitution of the matrices for the transformed process into Eq. (4.1) and showing that the joint density is not modified. As for PHDs similarity transformations may result in a description which is not a MAP since negative elements may appear in vector π, outside the diagonal of $\mathbf{D}_0$ or in matrix $\mathbf{D}_1$. These more general models are denoted as Rational Arrival Processes (RAPs) [5, 38] and are not considered here.

The equivalence relations presented for PHDs of different sizes can also be extended to MAPs. Let $(\pi, \mathbf{D}_0, \mathbf{D}_1)$ and $(\pi', \mathbf{D}_0', \mathbf{D}_1')$ be two MAPs of order m and n $(m > n)$, respectively. Let $\mathbf{V}$ be an $m \times n$ matrix with $\mathbf{V}\mathbb{1} = \mathbb{1}$, $\pi\mathbf{V} = \pi'$, $\mathbf{D}_0\mathbf{V} = \mathbf{V}\mathbf{D}_0'$ and $\mathbf{D}_1\mathbf{V} = \mathbf{V}\mathbf{D}_1'$, then both MAPs are equivalent (i.e. are different representations of the same stochastic process). Similarly, let $\mathbf{W}$ be an $n \times m$ matrix with $\mathbf{W}\mathbb{1} = \mathbb{1}$, $\boldsymbol{\pi} = \mathbf{W}\boldsymbol{\pi}'$, $\mathbf{W}\mathbf{D}_0 = \mathbf{D}_0'\mathbf{W}$ and $\mathbf{D}_1\mathbf{W} = \mathbf{D}_1'\mathbf{W}$, then both MAPs are equivalent. Again the equivalence of the representation can be proved by showing the equivalence of the joint densities [38]. For a MAP a minimal representation can be computed algorithmically [38] but the matrix representation might not and usually will not be a MAP, it is a RAP. The problem of finding a minimal MAP representation for a given MAP is still an unsolved problem.

An order n representation of a MAP contains $2n^2 - n$ parameters, if we assume that the initial vector $\boldsymbol{\pi}$ equals the embedded stationary vector $\boldsymbol{\pi}_s$. However, as shown in [155] usually a MAP of order n is characterized by only n^2 parameters which implies that the matrix representation is highly redundant and many equivalent representations even of the same order exist. This is a direct consequence of the equivalence relation using non-singular matrices $\mathbf{B}$. There are, of course, also equivalent representations of a larger order and there might be equivalent representations of a smaller order.

Example 4.4. We consider the following MAP with 3 states.

$$\boldsymbol{\pi} = [0.01961, 0.0, 0.98039],\ \mathbf{D}_0 = \begin{bmatrix} -1.00 & 0.0 & 0.0 \\ 89.1 & -100 & 0.0 \\ 0.00 & 0.00 & -100 \end{bmatrix},\ \mathbf{D}_1 = \begin{bmatrix} 0.50 & 0.0 & 0.50 \\ 0.55 & 0.0 & 10.35 \\ 1.00 & 0.0 & 99.00 \end{bmatrix}.$$

The initial vector $\boldsymbol{\pi}$ equals $\boldsymbol{\pi}_s$ the stationary vector after an event. The MAP is equivalent to the following MAP with only 2 states.

$$\boldsymbol{\pi}' = [0.019608, 0.980392],\ \mathbf{D}_0' = \begin{bmatrix} -1 & 0 \\ 0 & -100 \end{bmatrix},\ \mathbf{D}_1' = \begin{bmatrix} 0.5 & 0.5 \\ 1.0 & 99.0 \end{bmatrix}.$$

Equivalence is based on the relation $\mathbf{D}_0\mathbf{V} = \mathbf{V}\mathbf{D}_0'$, $\mathbf{D}_1\mathbf{V} = \mathbf{V}\mathbf{D}_1'$ and $\boldsymbol{\pi}' = \boldsymbol{\pi}\mathbf{V}$ with

$$\mathbf{V} = \begin{bmatrix} 1.0 & 0.0 \\ 0.9 & 0.1 \\ 0.0 & 1.0 \end{bmatrix}.$$

4.1.4 *MAPs as Counting Processes*

Rather than analyzing the process of inter-event times, one can alternatively consider the number of events in intervals $(0, t]$ which is denoted as the counting process of a MAP [2, 91, 120, 123]. Let $N(t)$ be the discrete random variable counting the number of events in $(0, t]$. If we assume $\boldsymbol{\pi} = \boldsymbol{\pi}_s$, then the first moment and the variance of the counting process are given by

$$
\begin{aligned}
E[N(t)] &= \frac{t}{E[X]} \text{ and} \\
VAR[N(t)] &= \frac{(E[X]+2)\,t}{(E[X])^2} \\
&- \frac{2\pi \mathbf{P}_s}{E[X]}\left(t\left(\frac{{}^T\pi(-\mathbf{D}_0)^{-1}}{E[X]} + \mathbf{Q}\right)^{-1} - \left(\mathbf{I} - e^{\mathbf{Q}t}\right)\left(\frac{{}^T\pi(-\mathbf{D}_0)^{-1}}{E[X]} + \mathbf{Q}\right)^{-2}\right)\mathbf{D}_1 \mathbb{1}
\end{aligned}
\tag{4.6}
$$

where $E[X] = \pi(-\mathbf{D}_0)^{-1}\mathbb{1}$, the first moment of the inter-event time, and $\mathbf{Q} = \mathbf{D}_0 + \mathbf{D}_1$. The first moment of the counting process can be easily derived from the first moment of the inter-event times, whereas the equations for higher order moments become complex which can be seen by considering the equation for the variance. The counting process can be described by a CTMC with an infinite state space, generator matrix

$$
\mathbf{Q}^c = \begin{bmatrix} \mathbf{D}_0 & \mathbf{D}_1 & \mathbf{0} & \cdots & \cdots \\ \mathbf{0} & \mathbf{D}_0 & \mathbf{D}_1 & \mathbf{0} & \cdots \\ \vdots & \ddots & \ddots & \ddots & \ddots \end{bmatrix} \tag{4.7}
$$

and initial vector $\pi^c = [\pi, 0, \ldots]$. Let $\psi^t = \pi^c e^{\mathbf{Q}^c t}$, then

$$
Prob(N(t) = k) = \psi^t_k \mathbb{1}
$$

where ψ^t_k is a vector built from the elements $kn+1, \ldots, (k+1)n$ of vector ψ^t and n is the dimension of the MAP. Vectors ψ^t_k can be computed from the following set of differential equations

$$
\begin{aligned}
\tfrac{d}{d\tau}\psi^\tau_0 &= \psi^\tau_0 \mathbf{D}_0 \\
\tfrac{d}{d\tau}\psi^\tau_k &= \psi^\tau_k \mathbf{D}_0 + \psi^\tau_{k-1}\mathbf{D}_1 \text{ for } k > 0
\end{aligned}
\tag{4.8}
$$

with initial conditions $\psi^0_0 = \pi$ and $\psi^0_k = \mathbf{0}$ for $k > 0$. To evaluate Eq. (4.8) using uniformization we define the vectors $\phi^{(0)}_k = \psi^0_k$ and use α, $\mathbf{P}_0$, $\mathbf{P}_1$ as in Sect. 4.1.2. Then define for $l > 0$

$$
\phi^{(l)}_k = \begin{cases} \phi^{(l-1)}_0 \mathbf{P}_0 & \text{if } k = 0, \\ \phi^{(l-1)}_k \mathbf{P}_0 + \phi^{(l-1)}_{k-1}\mathbf{P}_1 & \text{if } k > 0. \end{cases} \tag{4.9}
$$

With these vectors

$$
\psi^\tau_k = \sum_{l=0}^{\infty} \beta(\alpha\tau, l)\phi^{(l)}_k. \tag{4.10}
$$

For practical computation the infinite summation (parameter l) and the population (parameter k) are bounded such that for the computed vectors $1 - \sum_{k=0}^{k_{\max}} \psi_k^T < \epsilon$ for some small value ϵ holds.

4.2 MAPs of Order 2

One of the major problems in the application of MAPs is the highly redundant representation which makes the development of efficient methods for parameter fitting fairly complex. Thus, canonical representations based on a minimal number of parameters are very important. Unfortunately, such representations are not even known for the full class of PHDs, only for APHDs a canonical representation is available and can be easily computed (see Sect. 2.3.3). Since MAPs are more complex than PHDs, one cannot expect to find easily canonical representations for general MAPs, even for MAPs with an acyclic matrix $\mathbf{D}_0$, a canonical representation is still unknown. Therefore, some effort has been spend to analyze MAPs with a small number of states in detail to gain some more understanding, especially the case with only 2 states has been investigated in several papers [22, 69, 72, 73]. We briefly present the main results from [22] which are most general and can be used as a base for defining parameter fitting algorithms for MAPs of order 2.

If we assume that the initial vector of the MAP corresponds to the embedded stationary vector, a MAP of order 2 is completely characterized by setting 4 out of the 6 free parameters. Matrix $\mathbf{P}_s = (-\mathbf{D}_0)^{-1}\mathbf{D}_1$ is stochastic and has two eigenvalues $(1,\gamma)$ with $-1 \le \gamma < 1$. It can be shown [72] that

$$\rho_k = \frac{\gamma^k}{2}\left(\frac{E[X^2] - 2\cdot(E[X])^2}{E[X^2]-(E[X])^2}\right) = \frac{\gamma^k}{2}\left(\frac{2\left(\boldsymbol{\pi}(-\mathbf{D}_0)^{-2}\,\mathbb{1} - \left(\boldsymbol{\pi}(-\mathbf{D}_0)^{-1}\,\mathbb{1}\right)^2\right)}{2\boldsymbol{\pi}(-\mathbf{D}_0)^{-2}\,\mathbb{1} - \left(\boldsymbol{\pi}(-\mathbf{D}_0)^{-1}\,\mathbb{1}\right)^2}\right). \tag{4.11}$$

The equation shows that the coefficient of autocorrelation is geometrically decaying with parameter γ. A canonical representation can be computed based on 4 parameters [22], the two phase rates $\lambda(1), \lambda(2)$ with $0 < \lambda(1) \le \lambda(2)$ and $0 \le a, b < 1$. For $\gamma > 0$ the canonical form equals

$$\mathbf{D}_0 = \begin{bmatrix} -\lambda(1) & (1-a)\lambda(1) \\ 0 & -\lambda(2) \end{bmatrix}, \quad \mathbf{D}_1 = \begin{bmatrix} a\lambda(1) & 0 \\ (1-b)\lambda(2) & b\lambda(2) \end{bmatrix}. \tag{4.12}$$

In this case $\gamma = ab$, $\boldsymbol{\pi} = \left[\frac{1-b}{1-ab}, \frac{b-ab}{1-ab}\right]$ and $\mu_1 = \frac{(1-a)\lambda(1)+(1-b)\lambda(2)}{\lambda(1)\lambda(2)(1-ab)}$. The remaining quantities, like higher order moments or joint moments can be easily computed from the representation.

For $\gamma < 0, 0 \le a \le 1, 0 < b \le 1$ and if $a = 1$ $\lambda(1) \ne \lambda(2)$ are required. The canonical representation equals in this case

$$\mathbf{D}_0 = \begin{bmatrix} -\lambda(1) & (1-a)\lambda(1) \\ 0 & -\lambda(2) \end{bmatrix}, \ \mathbf{D}_1 = \begin{bmatrix} 0 & a\lambda(1) \\ b\lambda(2) & (1-b)\lambda(2) \end{bmatrix}. \quad (4.13)$$

We obtain then $\gamma = -ab$, $\boldsymbol{\pi} = \left[\frac{b}{1+ab}, 1 - \frac{b}{1+ab}\right]$ and $\mu_1 = \frac{b\lambda(2)+\lambda(1)}{(1+ab)\lambda(1)\lambda(2)}$.

Both canonical representations are characterized by an acyclic matrix $\mathbf{D}_0$. Since we have two states, for the squared coefficient of variation $0.5 \le C^2$ has to hold. Of course, the value of C^2 determines the flexibility of the MAP. For C^2 close to 0.5, $\mathbf{D}_0$ tends to the matrix of an Erlang 2 distribution with a small probability to leave the distribution after phase one. This implies that γ tends to 0 in this case.

If the matrix $\mathbf{D}_0$ is restricted to a diagonal matrix, the marginal distribution of the MAP becomes hyper-exponential which implies that $C^2 \ge 1$. The boundaries for the reachable three moments and reachable parameters γ have been investigated in [69].

Example 4.5. We consider as an example the MMPP from Example 4.2. The MMPP representation is not in canonical form. Since the eigenvalues of matrix $\mathbf{D}_0$ are $(1, 0.9434)$, the process can be represented in the first canonical form 4.12. The representation equals

$$\boldsymbol{\pi} = [0.17587, 0.82413],\ \mathbf{D}_0 = \begin{bmatrix} -1.04945 & 0.04945 \\ 0.00000 & -10.10055 \end{bmatrix} \text{ and } \mathbf{D}_1 = \begin{bmatrix} 1.00000 & 0.00000 \\ 0.10055 & 10.00000 \end{bmatrix}.$$

Since $E[T] = 0.25$ and $E[T^2] = 0.33726$, the lag k autocorrelation coefficient equals $\rho_k = (0.9434)^k \cdot 0.38627$.

4.3 BMAPs and MMAPs

MAPs generate a single type of events. This can be extended by allowing K different event types resulting in a Marked MAP (MMAP) which is defined as $(\boldsymbol{\pi}, \mathbf{D}_0, \mathbf{D}_1, \ldots, \mathbf{D}_K)$ where $(\boldsymbol{\pi}, \mathbf{D}_0, \sum_{k=1}^K \mathbf{D}_k)$ is a MAP and all matrices $\mathbf{D}_k$ are non-negative. MMAPs have been originally introduced in [65]. If one interprets the different arrivals as batches of arrivals (i.e. matrix $\mathbf{D}_k$ contains transition rates that are accompanied by the arrival of a batch of k events), one arrives at a Batch MAP (BMAP) [113, 114]. Since BMAPs and MMAPs only differ in the interpretation of events, we consider MMAPs here. As before, we assume $\boldsymbol{\pi} = \boldsymbol{\pi}_s$ where $\boldsymbol{\pi}_s$ the unique solution of $\boldsymbol{\pi}_s(-\mathbf{D}_0)^{-1}\sum_{k=1}^K \mathbf{D}_k = \boldsymbol{\pi}_s$ subject to $\boldsymbol{\pi}_s \mathbb{1} = 1$.

It is easy to define a subclass of MMAPs where inter-event times of all event types are independently and identically distributed. These are distributions generating different event types. In this case $\mathbf{D}_k = \mathbf{d}_k\boldsymbol{\pi}$ for some non-negative column vector $\mathbf{d}_k$.

The inter-event time distribution of an MMAP is a PHD $(\boldsymbol{\pi}_s, \mathbf{D}_0)$. If we consider the time between two events of type k, then the stationary distribution immediately after a type k event is given by the solution of $\boldsymbol{\pi}_k\left(-\left(\mathbf{D}_0 + \sum_{l=1, l\neq k}^K \mathbf{D}_l\right)\right)^{-1}\mathbf{D}_k = \boldsymbol{\pi}_k$

subject to $\boldsymbol{\pi}_k \mathbb{1} = 1$. The inverse matrix exists since matrix $\mathbf{D}_0 + \sum_{k=1}^{K} \mathbf{D}_k$ is irreducible which follows from the conditions required for MAPs. The inter-event time distribution between two occurrences of type k events then has a PHD $(\boldsymbol{\pi}_k, \mathbf{D}_0 + \sum_{l=1, l\neq k}^{K} \mathbf{D}_l)$. Specifically for BMAPs, the mean rate of arriving events can be computed as

$$\boldsymbol{\psi} \sum_{k=1}^{K} k\mathbf{D}_k \mathbb{1} \text{ where } \boldsymbol{\psi}\left(\mathbf{D}_0 + \sum_{k=1}^{K} \mathbf{D}_k\right) = 0 \text{ and } \boldsymbol{\psi}\mathbb{1} = 1. \tag{4.14}$$

Finally, characteristics that describe the correlation between different events and event types are introduced [39, 65]. The joint density of the event process is based on the inter-event times and the type of the arriving events and equals

$$f(x_1, k_1, x_2, k_2, \ldots, x_l, k_l) = \boldsymbol{\pi} e^{\mathbf{D}_0 x_1} \mathbf{D}_{k_1} e^{\mathbf{D}_0 x_2} \mathbf{D}_{k_2} \ldots e^{\mathbf{D}_0 x_l} \mathbf{D}_{k_l} \mathbb{1} \tag{4.15}$$

for $x_i \geq 0$ and $k_i \in \{1, \ldots, K\}$ $(i = 1, \ldots, l)$. The probability of observing a type k event under the condition that an event is observed equals

$$p_k^{arr} = \boldsymbol{\pi}(-\mathbf{D}_0)^{-1} \mathbf{D}_k \mathbb{1}. \tag{4.16}$$

Similarly, the probability of observing a sequence $k_1, \ldots, k_l$ of event types equals

$$p_{k_1,\ldots,k_l}^{arr} = \boldsymbol{\pi} \left(\prod_{i=1}^{l} (-\mathbf{D}_0)^{-1} \mathbf{D}_{k_i} \right) \mathbb{1}. \tag{4.17}$$

Define $E[X_0^i, k, X_1^j]$ as the joint moment of order i, j of two consecutive events under the condition that the first event is of type k. This measure can be computed as (see [39])

$$E[X_0^i, k, X_1^j] = \frac{i!j!}{p_k^{arr}} \boldsymbol{\pi}(-\mathbf{D}_0)^{-(i+1)} \mathbf{D}_k (-\mathbf{D}_0)^{-j} \mathbb{1}. \tag{4.18}$$

Example 4.6. Consider the following MMAP

$$\boldsymbol{\pi} = [0.5, 0.5],\ \mathbf{D}_0 = \begin{bmatrix} -1 & 0 \\ 0 & -2 \end{bmatrix}, \mathbf{D}_1 = \begin{bmatrix} 0 & 1 \\ 0 & 0 \end{bmatrix},\ \mathbf{D}_2 = \begin{bmatrix} 0 & 0 \\ 2 & 0 \end{bmatrix}$$

that generates an alternating sequence of type 1 and 2 events. Inter-event times are exponentially distributed with mean 1 if the last event was of type 2 and with mean 0.5 if the last event was of type 1. An arrival is with probability $p_1^{arr} = 1/2$ of type 1 and with probability $p_2^{arr} = 1/2$ of type 2.

The equivalence of MAPs can be easily extended to MMAPs by simply extending the required conditions to all matrices $\mathbf{D}_k$ as done in [38]. As in the case with one event type, the construction allows the generation of representations that are non-Markovian and are denoted as Marked RAPs (MRAPs). Almost nothing is known about canonical representations of MMAPs, even the simplest case with only 2 states and 2 classes has not been analyzed yet.

4.4 Properties

The family of MAPs is closed under several operations such as superposition and random thinning. We first refer to the results discussed in [115] where the authors stated not only that a PH-renewal process is a MAP but also that a sequence of PH distributed interarrival times selected due to a Markov process is also a MAP. The latter one was originally studied in [102].

Note that a PH-renewal process is a PHD where after hitting the absorbing state the process is restarted, i.e. the new initial state is chosen according to a PHD. The formal description of the resulting MAP is given in Sect. 4.1.1. However, at an arrival the new initial phase is selected according to the initial distribution vector of the PHD. Thus the new phase is independent of the past and there is no possibility to define correlated patterns.

On the other hand if one aims to obtain a sequence of distinct PHD interarrival times the resulting process is a MAP. In this case, the successive interarrival times are selected according to the certain Markov chain with transition matrix $\mathbf{P}$. Let $PH_A = (\boldsymbol{\pi}^{(A)}, \mathbf{D}_0^{(A)})$ be of order n, and $PH_B = (\boldsymbol{\pi}^{(B)}, \mathbf{D}_0^{(B)})$ be of order m. Then the MAP with the described behavior is given in Eq. (4.19).

$$\mathbf{D}_0 = \begin{bmatrix} \mathbf{D}_0^{(A)} & \mathbf{0} \\ \mathbf{0} & \mathbf{D}_0^{(B)} \end{bmatrix} \text{ and } \mathbf{D}_1 = \begin{bmatrix} \mathbf{P}(1,1)\mathbf{d}_1^{(A)}\boldsymbol{\pi}^{(A)} & \mathbf{P}(1,2)\mathbf{d}_1^{(A)}\boldsymbol{\pi}^{(B)} \\ \mathbf{P}(2,1)\mathbf{d}_1^{(B)}\boldsymbol{\pi}^{(A)} & \mathbf{P}(2,2)\mathbf{d}_1^{(B)}\boldsymbol{\pi}^{(B)} \end{bmatrix}. \tag{4.19}$$

One can see that no correlation between alternating PHDs is decoded here. It is in principle possible to substitute, e.g. the entry $\mathbf{d}_1^{(A)}\boldsymbol{\pi}^{(A)}$, by a valid $\mathbf{D}_1^{(A)}$ matrix to put the autocorrelation of the first MAP into consideration.

Since the family of MAPs is closed under superposition, we summarize this operation in the following. Let $(\boldsymbol{\pi}^{(A)}, \mathbf{D}_0^{(A)}, \mathbf{D}_1^{(A)})$, and $(\boldsymbol{\pi}^{(B)}, \mathbf{D}_0^{(B)}, \mathbf{D}_1^{(B)})$ be two independent MAPs. Then the superposition of the two MAPs is also a MAP with representation given in Eq. (4.20).

$$\mathbf{D}_0^{(C)} = \mathbf{D}_0^{(A)} \oplus \mathbf{D}_0^{(B)} \text{ and } \mathbf{D}_1^{(C)} = \mathbf{D}_1^{(A)} \oplus \mathbf{D}_1^{(B)}. \tag{4.20}$$

Note that more than 2 MAPs can be used in the superposition construction.

As a numerical example we consider the superposition of two identical IPPs defined in Fig. 4.2. The resulting MAP is visualized in Fig. 4.3.

a

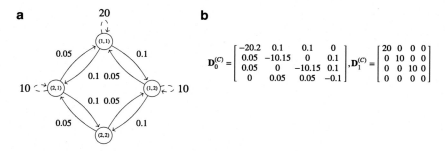

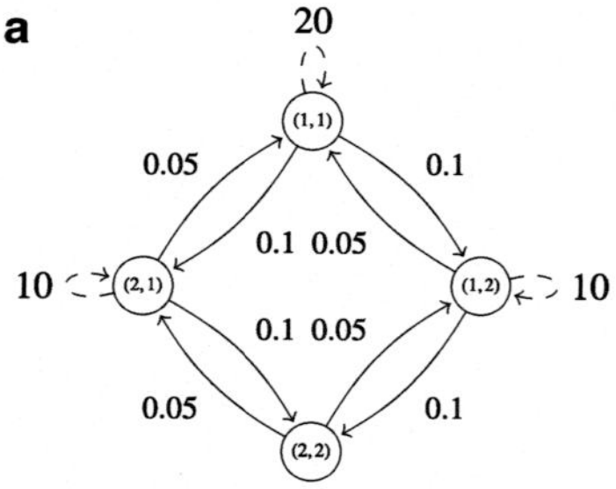

b

$$\mathbf{D}_0^{(C)} = \begin{bmatrix} -20.2 & 0.1 & 0.1 & 0 \\ 0.05 & -10.15 & 0 & 0.1 \\ 0.05 & 0 & -10.15 & 0.1 \\ 0 & 0.05 & 0.05 & -0.1 \end{bmatrix}, \mathbf{D}_1^{(C)} = \begin{bmatrix} 20 & 0 & 0 & 0 \\ 0 & 10 & 0 & 0 \\ 0 & 0 & 10 & 0 \\ 0 & 0 & 0 & 0 \end{bmatrix}$$

Fig. 4.3 The superposition of two identical IPPs introduced in Fig. 4.2 (**a**) Diagrammatic representation of the superposition process (**b**) The matrices of the MAP resulting from superposition

4.5 Concluding Remarks

MAPs have been introduced by Neuts [124], the matrix representation of MAPs, which is used here and in most other papers on the subject, is due to Lucantoni [113]. From [155] the number of free parameters of a MAP is known and [22] establishes a canonical form of MAPs of order 2. However, the theoretical foundation of MAPs is less advanced than the basis of PHDs which is not surprising since the future behavior of MAP may depend on the whole history and not only on the time since the last event occurred as it is the case for PHDs.

Several extensions of MAPs exist. Foremost the extension to generate different arrival types or batches of arrival which are denoted as MMAPs [65] or BMAPs [113]. Both process types are useful in practice but are even more complex than MAPs. The class of Markov additive processes [4], sometimes also abbreviated as MAP, is even more general and includes BMAPs or MMAPs. Another generalization are Rational Arrival Processes (RAPs) [5] which result from a linear algebraic view without probabilistic interpretation similar to Matrix Exponential distributions. These processes are rarely used yet, since the theory is not completely developed although some newer results show interesting relations between MAPs and RAPs [37].

Chapter 5
Parameter Fitting of MAPs

Fitting the parameters of a MAP is much more complex than the parameter fitting for PHDs. The major reasons for the complexity of the fitting problem are missing canonical representations for MAPs and the necessity to consider long traces to adequately capture the correlation. Although most algorithms for PHDs, that have been presented in Chap. 3, can be extended for MAPs, the effort for finding MAP parameters is usually much higher and the algorithms are less reliable and stable when applied to MAPs. Nevertheless, first approaches for the parameterization are available nowadays and in many situations these approaches can be applied for practical problems.

This chapter gives an overview of available methods for the parameter fitting of MAPs. As for PHDs one can distinguish between approaches that compute the parameters according to some derived measures, like joint moments or the coefficients of autocorrelation, and methods that consider the whole trace for parameter fitting. For MAPs different methods can be mixed because the parameters in $\mathbf{D}_0$ and $\mathbf{D}_1$ may be computed separately in subsequent steps.

In the first section, the available approaches for MAP parameter fitting based on moments and joint moments are introduced. Afterwards, EM algorithms are extended from PHDs to MAPs. Then, Sect. 5.3 shows different combinations of fitting techniques that separately fit the parameters of the two MAP matrices. The final section of this chapter is devoted to a brief overview of EM algorithms that compute the MAP parameters according to measures of the counting rather than the inter-event time process.

5.1 Moment and Joint Moment Based Fitting

Available methods to compute the parameters of a MAP with respect to the moments and joint moments are limited to relatively simple MAP structures to keep the problem feasible. In principle, one can write down the equations for higher order

P. Buchholz et al., *Input Modeling with Phase-Type Distributions and Markov Models: Theory and Applications*, SpringerBriefs in Mathematics,
DOI 10.1007/978-3-319-06674-5_5, © Peter Buchholz, Jan Kriege, Iryna Felko 2014

Table 5.1 Bounds for the Hankel determinants of a $MAP(2)$

$h_1 > 0$	Hypo-exponential ($C^2 \leq 1$)	Hyper-exponential ($C^2 \geq 1$)
h_2	$-0.25 \leq h_2 < 0$	$0 < h_2$
h_3	$h_2\left(1 - h_2 - 2\sqrt{-h_2}\right) \leq h_3 \leq -h_2^2$	$0 < h_3$

moments or joint moments (cf. Sect. 4.1.2) and compute the MAP parameters from the inverted equations. Unfortunately, the resulting equations are non-linear and available solution and optimization methods do not work for the general problems. Here we present first approaches for MAPs with two states and introduce afterwards an approach that composes MAPs with two states for an improved moment and autocorrelation fitting. Additional approaches can be found in Sect. 5.3 where in two steps the distribution and the correlation are considered for parameter fitting.

5.1.1 Parameter Fitting for MAPs with Two States

The parameter fitting for MAPs with two states can exploit the knowledge of the canonical representation (cf. Sect. 4.2). As already shown, a two state MAP is characterized by four parameters, namely the two rates $\lambda(1)$ and $\lambda(2)$ and the two values a and b. This implies that, at least in principle, four measures of a trace can be used to fit the MAP parameters such that the resulting MAP reflects exactly these measures. Commonly the first three moments μ_1, μ_2, μ_3, and the parameter γ are used as measures. Since ρ_k is proportional to γ^k [cf. Eq. (4.11)], one coefficient of autocorrelation can be reproduced, usually $k = 1$ is chosen in this case.

The four free parameters of a MAP of order 2 have to observe several restrictions such that only a limited subset of possible combinations of the first three moments and the coefficient of autocorrelation can be reached. Obviously, $1.5 \leq \mu_2/\mu_1$ has to hold. If μ_2 reaches the lower boundary, the MAP corresponds to an Erlang 2 distribution without any flexibility in choosing μ_3 and γ.

The fitting procedure presented in [22] uses the Hankel determinants rather than the moments. The Hankel determinants are defined as

$$h_1 = \mu_1,\ h_2 = \frac{\mu_2}{2\mu_1^2} = \frac{C^2 - 1}{2} \text{ and } h_3 = \frac{\mu_3}{6\mu_1^3} - \frac{\mu_2^2}{4\mu_1^4}. \tag{5.1}$$

The reachable Hankel determinants are shown in Table 5.1 [22]. Table 5.2 contains the bounds for parameter γ depending on the Hankel determinants. For $h_2 = 0$ the process degenerates to an exponential distribution.

From the first three moments (or Hankel determinants) the parameters $\lambda(i)$ $(i = 1, 2)$ and p, which equals the probability of entering the resulting APHD in state 1, can be computed. If $h_2 \geq 0$, which implies $C^2 \geq 1$, the values are obtained from

Table 5.2 Bounds for the parameter γ in terms of the Hankel determinants

Area	Condition	Lower bound	Upper bound
A	$h_2 < 0$	$-\left(\frac{h_3}{h_2}+h_2\right)$	$-\frac{\left(\sqrt{-h_3}+h_2\right)^2}{h_2}$
B	$h_2 > 0 \;\&\; \frac{h_3}{h_2}+h_2 < 1$	$-\left(\frac{h_3}{h_2}+h_2\right)$	1
C	$h_2 > 0 \& 1 \leq \frac{h_3}{h_2}+h_2$	$\frac{h_3+h_2^2-h_2-\sqrt{\left(h_3+h_2^2-h_2\right)^2+4h_2^3}}{h_3+h_2^2-h_2+\sqrt{\left(h_3+h_2^2-h_2\right)^2+4h_2^3}}$	1

$$
\begin{aligned}
\lambda(i) &= \frac{h_3+h_2^2+h_2 \pm \sqrt{(h_3+h_2^2+h_2)^2+4h_2h_3}}{2h_1h_3} \qquad (i=1,2) \\
p &= \frac{-h_3-h_2^2+h_2+\sqrt{(h_3+h_2^2+h_2)^2-4h_2h_3}}{h_3+h_2^2+h_2+\sqrt{(h_3+h_2^2+h_2)^2-4h_2h_3}}, \; \alpha = \frac{\lambda(1)}{\lambda(2)}.
\end{aligned} \tag{5.2}
$$

If $h_2 < 0$, which implies $C^2 < 1$, we obtain

$$
\begin{aligned}
\lambda(i) &= \frac{h_3+h_2^3+h_2 \pm \sqrt{(h_3-h_2^2+h_2)^2+4h_2h_3}}{2h_1h_3} \qquad (i=1,2) \\
p &= \frac{h_3+h_2^2-h_2+\sqrt{(h_3+h_2^2+h_2)^2-4h_2h_3}}{-h_3-h_2^2-h_2+\sqrt{(h_3+h_2^2+h_2)^2-4h_2h_3}}, \; \alpha = \frac{\lambda(1)}{\lambda(2)}.
\end{aligned} \tag{5.3}
$$

Furthermore, $\lambda(1) \leq \lambda(2)$ is required in the canonical representation such that $\alpha \leq 1$ follows. The three parameters define an APHD in the canonical form presented in Eq. (4.12) or (4.13).

The coefficient of autocorrelation is determined by γ which equals ab or $-ab$ and the first two moments (cf. Sect. 4.2). For a given $\gamma \geq 0$ the parameters a and b are given by

$$
\begin{aligned}
a &= \tfrac{1}{2\alpha}\left(1+\alpha\gamma-p(1-\gamma)-\sqrt{(1+\alpha\gamma-p(1-\gamma))^2-4\alpha\gamma}\right), \\
b &= \tfrac{1}{2}\left(1+\alpha\gamma-p(1-\gamma)+\sqrt{(1+\alpha\gamma-p(1-\gamma))^2-4\alpha\gamma}\right).
\end{aligned} \tag{5.4}
$$

For $\gamma \leq 0$ the parameters are given by

$$
a = \frac{-\gamma}{p(1-\gamma)-\alpha\gamma}, \quad b = p(1-\gamma)-\alpha\gamma. \tag{5.5}
$$

For feasible values, the equations result in a valid MAP of order 2. However, for values outside the feasible region, the computations may fail or result in MAP parameters outside the valid region. If this is the case, then the trace measures cannot be exactly represented by a MAP with only 2 states. One may then use a MAP with more states. The parameters of a such a MAP have to be fitted with one

Table 5.3 Moments and lag 1 coefficient of autocorrelation for the traces

Measure	lbl-trace	pAug-trace	tudo-trace
$\hat{\mu}_1$	1.000	1.000	1.000
$\hat{\mu}_2$	2.942	4.223	160.5
$\hat{\mu}_3$	16.84	64.76	178498
$\hat{\rho}_1$	0.155	0.200	0.562

of the approaches presented in the following sections. Alternatively, one can try to find a MAP of order 2 that approximates the measures from the trace as close as possible. The difference between the measures of the trace and the MAP is usually measured in terms of the sum of the squared difference between the moments and coefficients of autocorrelation of the MAP and the trace. This measure is also used in approximate moment fitting for PHDs (cf. Sect. 3.2.2) and approximate joint moment fitting for MAPs (cf. Sect. 5.3). As shown in [24] the use of general optimization algorithms to minimize the squared difference between the first three moments and the lag 1 coefficient of autocorrelation of the trace and the MAP yields poor results since the feasible region of the parameters is non-convex. In [24] an alternative optimization approach is presented which utilizes the knowledge of the surface of the feasible region. A simple alternative to this optimization approach is the ordered moment adjusting method which is also presented in [24]. In this case, an order is defined for the measures. μ_1, which can always be achieved by a $MAP(2)$, obtains the highest priority. Usually μ_2 is the second parameter, followed by μ_3 or γ. Then the parameters of the MAP are chosen to match the measures in the given order. If a measure cannot be exactly matched by the MAP, the parameter is set to the nearest value which can be reached by the MAP.

Example 5.1. We consider 3 common traces, namely the traces *lbl* and *pAug* from the Internet traffic archive [84] and the trace *tudo* which contains measurements from the Web server of the TU Dortmund. For all three traces the mean values are scaled to 1 which does not alter the fitting problem because the first moment can be easily scaled by a multiplicative constant. Table 5.3 contains the first three moments and the lag 1 coefficient of autocorrelation for the traces.

For the traces *lbl* and *pAug* the first three moments and the lag 1 coefficient of autocorrelation can be exactly matched by a MAP with only 2 states. For the trace *tudo*, the first 3 moments can be matched with a 2 state MAP, but it is impossible to match the lag 1 coefficient of autocorrelation, even if the second and third moment are not matched. More phases are required to reach ρ_1 above 0.5. The following MAPs result from the traces.

$$\text{lbl-trace} \quad \mathbf{D}_0 = \begin{bmatrix} -0.45017 & 0.13804 \\ 0.00000 & -1.62770 \end{bmatrix}, \mathbf{D}_1 = \begin{bmatrix} 0.31213 & 0.00000 \\ 0.12597 & 1.50173 \end{bmatrix}$$
$$\text{pAug-trace} \quad \mathbf{D}_0 = \begin{bmatrix} -0.14297 & 0.05904 \\ 0.00000 & -1.22771 \end{bmatrix}, \mathbf{D}_1 = \begin{bmatrix} 0.08393 & 0.00000 \\ 0.01468 & 1.21303 \end{bmatrix}$$
$$\text{tudo-trace} \quad \mathbf{D}_0 = \begin{bmatrix} -0.00268 & 0.00003 \\ 0.00000 & -1.27023 \end{bmatrix}, \mathbf{D}_1 = \begin{bmatrix} 0.00265 & 0.00000 \\ 0.00001 & 1.27023 \end{bmatrix}$$

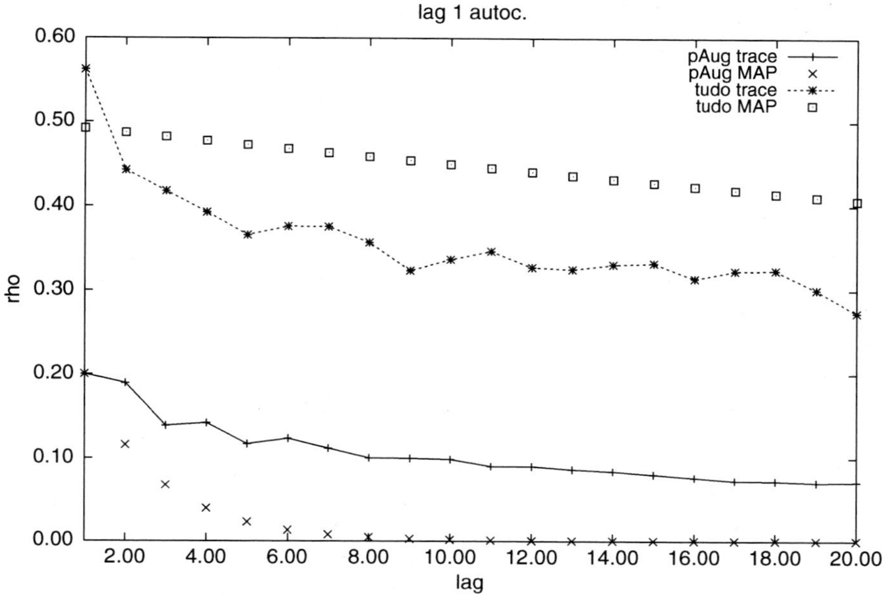

Fig. 5.1 Lag k coefficient of autocorrelation for the traces *pAug* and *tudo*

The MAP for the *tudo*-trace has a lag 1 coefficient of autocorrelation of 0.491, a value below 0.562, which is the value of the trace.

Even if ρ_1 is matched exactly, this does not imply that the higher lags are also adequately captured by the MAP. Figure 5.1 shows the course of ρ_k for the trace *pAug* and *tudo* and the fitted MAPs. It can be seen that for the trace *pAug*, the lag k coefficient of the fitted MAP vanishes quickly whereas for *tudo* the lag k coefficients of the fitted MAP are larger than the values of the trace for $k \geq 2$.

5.1.2 A Compositional Approach

MAPs of order 2 have the advantage that a canonical form is available for them and parameters can be computed analytically if the measures to be matched by the MAP fall into the feasible region. However, the feasible region is often too small for practical problems and it is also not sufficient to consider only ρ_1 and not higher lags. An approach that combines, up to some extend, the fitting of MAPs of a small order with the flexibility of larger MAPs, is the *Kronecker Product Composition* (KPC) of MAPs [42]. In this approach, MAPs are composed from smaller MAPs using Kronecker products. Let $(\mathbf{D}_0^{(1)}, \mathbf{D}_1^{(1)}), \ldots, (\mathbf{D}_0^{(J)}, \mathbf{D}_1^{(J)})$ be a set of J MAPs, where the matrices $\mathbf{D}_0^{(2)}, \ldots, \mathbf{D}_0^{(J)}$ are diagonal matrices with negative diagonal entries.The KPC of these MAPs is a MAP $(\mathbf{D}_0^{KPC}, \mathbf{D}_1^{KPC})$ with

$$\mathbf{D}_0^{KPC} = (-1)^{J-1}(\mathbf{D}_0^{(1)} \otimes \mathbf{D}_0^{(2)} \otimes \ldots \otimes \mathbf{D}_0^{(J)}),\ \mathbf{D}_1^{KPC} = \mathbf{D}_1^{(1)} \otimes \mathbf{D}_1^{(2)} \otimes \ldots \otimes \mathbf{D}_1^{(J)}. \quad (5.6)$$

If $n^{(j)}$ is the order of MAP $(\mathbf{D}_0^{(j)}, \mathbf{D}_1^{(j)})$, then the composed MAP has order $n = \prod_{j=1}^{J} n^{(j)}$. Although the MAPs with the numbers 2 through J are restricted to diagonal matrices $\mathbf{D}_0$, the method is general since the first MAP can be arbitrary.

The KPC has the property that measures of the composed MAP can be derived easily from measures of the composing MAPs. We consider the results here for the KPC of two MAPs which can be easily generalized because the composition of J MAPs is equivalent to $J-1$ compositions of 2 MAPs. Let $(\mathbf{D}_0^{(0)}, \mathbf{D}_1^{(0)}) = (-\mathbf{D}_0^{(1)} \otimes \mathbf{D}_0^{(2)}, \mathbf{D}_1^{(1)} \otimes \mathbf{D}_1^{(2)})$ be the KPC of two MAPs, where $\mathbf{D}_0^{(2)}$ is a diagonal matrix. Let $\mu_k^{(i)}$, $C^{(i)2}$, $\mu_{kl}^{(i)}$ and $\rho_k^{(i)}$ $(i = 0, 1, 2)$ be the moments, squared coefficient of variation, joint moments and lag k coefficients of autocorrelation. The measures of the composed MAP are then computed as

$$\mu_k^{(0)} = \frac{\mu_k^{(1)} \mu_k^{(2)}}{k!}, \qquad \mu_{kl}^{(0)} = \frac{\mu_{kl}^{(1)} \mu_{kl}^{(2)}}{k!l!},$$
$$1 + C^{(0)2} = \frac{(1 + C^{(1)2})(1 + C^{(2)2})}{2},\ C^{(0)2}\rho_k^{(0)} = C^{(1)2}\rho_k^{(1)} + C^{(2)2}\rho_k^{(2)} + C^{(1)2}C^{(2)2}\rho_k^{(1)}\rho_k^{(2)}. \quad (5.7)$$

Equation 5.7 can now be used as base for a fitting algorithm. The first MAP can be arbitrary, whereas the remaining MAPs $2, \ldots, J$ have to be hyper-exponential, which implies $C^{(j)2} \geq 1$ for $j = 2, \ldots, J$. Usually, the approach composes MAPs of order 2 because for these MAPs, the boundaries of the feasible regions are known. If the coefficient of variation of the trace is not too small, the first MAP can be chosen as MAP of order 2 in canonical form and the remaining MAPs are hyper-exponential MAPs of order 2. In this case, feasible values for the first three moments and parameter γ for each MAP are known such that a non-linear optimization algorithm can be applied to compute the parameters of the MAPs.

5.2 Trace Based Fitting of MAPs

Parameter fitting of MAPs according to the values in a trace $\mathcal{T}$ is done with EM algorithms which are extensions of the EM algorithms for the parameter fitting of PHDs (cf. Sect. 3.1.2). We use the same notation as for PHDs. However, since the MAP has to consider the correlation structure, the likelihood function cannot be defined separately for every trace entry, it has to consider the whole trace such that Eq. (3.8) becomes

$$\mathcal{L}((\mathbf{D}_0, \mathbf{D}_1)|\mathcal{T}) = \boldsymbol{\pi} \prod_{i=1}^{m} e^{\mathbf{D}_0 t_i} \mathbf{D}_1 \mathbb{1}. \quad (5.8)$$

Thus, the optimization problem becomes

$$(\mathbf{D}_0,\mathbf{D}_1)^* = \arg\max_{(\mathbf{D}_0,\mathbf{D}_1)} \boldsymbol{\pi} \prod_{i=1}^{m} e^{\mathbf{D}_0 t_i} \mathbf{D}_1 \mathbb{1}, \tag{5.9}$$

with the additional constraint that $\mathbf{D}_0 + \mathbf{D}_1$ is an irreducible generator matrix. Similar to Sect. 3.1.2 some quantities are defined. B_i is the number of times the process starts in state i, the total time spent in state i is given by Z_i, and the total observed number of jumps from state i to state j without generating an event is N_{ij}, for $i \neq j$, and M_{ij} is the total observed number of jumps from state i to j ($i, j \in \mathcal{S}$) when generating an event. In terms of these quantities the likelihood function can be expressed as

$$\begin{aligned}
&\mathcal{L}((\mathbf{D}_0,\mathbf{D}_1)|\mathcal{T}) \\
&= \prod_{i=1}^{n} \pi(i)^{B_i} \prod_{i=1}^{n} e^{Z_i \mathbf{D}_0(i,i)} \prod_{i=1}^{n} \prod_{j=1,j\neq i}^{n} \mathbf{D}_0(i,j)^{N_{ij}} \prod_{i=1}^{n} \prod_{j=1,j\neq i}^{n} \mathbf{D}_1(i,j)^{M_{ij}} \mathbb{1}.
\end{aligned} \tag{5.10}$$

Based on Eq. (5.10), an EM algorithm can be formulated, several variants can be found in the literature [25, 31, 97]. We present here a variant which is an extension of Algorithm 1. The approach can also be extended to BMAPs or MMAPs.

Since for MAPs the complete arrival sequence of arrivals and not only single arrivals have to be considered the notation has to be extended. Let $\mathbf{ff}^{(k)}_{(\mathbf{D}_0,\mathbf{D}_1)}$ and $\mathbf{bb}^{(k)}_{(\mathbf{D}_0,\mathbf{D}_1)}$ be the forward and backward vector before and after the kth event ($k = 1,\ldots,m$). The vectors are defined as

$$\begin{aligned}
\mathbf{ff}^{(k)}_{(\mathbf{D}_0,\mathbf{D}_1)} &= \begin{cases} \boldsymbol{\pi} & \text{if } k = 1 \\ \mathbf{ff}^{(k-1)}_{(\mathbf{D}_0,\mathbf{D}_1)} e^{\mathbf{D}_0 t_{k-1}} \mathbf{D}_1 & \text{if } 1 < k \leq m \end{cases} \\
\mathbf{bb}^{(k)}_{(\mathbf{D}_0,\mathbf{D}_1)} &= \begin{cases} \mathbf{D}_1 \mathbb{1} & \text{if } k = m \\ \mathbf{D}_1 e^{\mathbf{D}_0 t_{k+1}} \mathbf{bb}^{(k+1)}_{(\mathbf{D}_0,\mathbf{D}_1)} & \text{if } 1 \leq k < m \end{cases}
\end{aligned} \tag{5.11}$$

Backward and forward vectors and matrices for the kth event in the trace are given by

$$\begin{aligned}
&\mathbf{f}^{(k)}_{(\mathbf{D}_0,\mathbf{D}_1),t} = \mathbf{ff}^{(k)}_{(\mathbf{D}_0,\mathbf{D}_1)} e^{\mathbf{D}_0 t}, \ \mathbf{b}^{(k)}_{(\mathbf{D}_0,\mathbf{D}_1),t} = e^{\mathbf{D}_0 t} \mathbf{bb}^{(k)}_{(\mathbf{D}_0,\mathbf{D}_1)}, \text{ and} \\
&\mathbf{F}^{(k)}_{(\mathbf{D}_0,\mathbf{D}_1),t} = \int_0^t \left(\mathbf{f}^{(k)}_{(\mathbf{D}_0,\mathbf{D}_1),t-u}\right)^T \left(\mathbf{b}^{(k)}_{(\mathbf{D}_0,\mathbf{D}_1),u}\right)^T du.
\end{aligned} \tag{5.12}$$

The vectors and matrix can alternatively be computed using uniformization. The resulting computations are similar to those presented in Sect. 3.1.2. The E-step of the algorithm uses the computed vectors and matrices.

Algorithm 4 EM algorithm for general MAPs

Input: Trace data $\mathcal{T} = t_1, \ldots, t_m$;
Output: MAP $(\mathbf{D}_0, \mathbf{D}_1)$;
1: Choose MAP $(\mathbf{D}_0^{(0)}, \mathbf{D}_1^{(0)})$ and set $r = 0$;
2: **repeat**
3: $\mathbf{ff}^{(1)} = \boldsymbol{\pi}$;
4: **for** $i = 2 \to m$ **do**
5: Compute and store $\mathbf{ff}^{(i)}_{(\mathbf{D}_0^{(r)}, \mathbf{D}_1^{(r)}), t_i}$ using Eq. (5.11) ;
6: **end for**
7: **for** $i = m \to 1$ **do**
8: Compute $\mathbf{bb}^{(i)}_{(\mathbf{D}_0^{(r)}, \mathbf{D}_1^{(r)}), t_i}$, $\mathbf{b}^{(i)}_{(\mathbf{D}_0^{(r)}, \mathbf{D}_1^{(r)}), t_i}$, $\mathbf{f}^{(i)}_{(\mathbf{D}_0^{(r)}, \mathbf{D}_1^{(r)}), t_i}$ and $\mathbf{F}^{(i)}_{(\mathbf{D}_0^{(r)}, \mathbf{D}_1^{(r)}), t_i}$ using Eq. (5.11) and Eq. (5.12)
9: **end for**
10: **E-step**: Compute the conditional expectations using Eq. (5.13);
11: **M-step**: Compute $(\mathbf{D}_0^{(r+1)}, \mathbf{D}_1^{(r+1)})$ using Eq. (5.14) and set $r = r + 1$;
12: **until** $\|\mathbf{D}_0^{(r)} - \mathbf{D}_0^{(r-1)}\| + \|\mathbf{D}_1^{(r)} - \mathbf{D}_1^{(r-1)}\| < \epsilon$;
13: **return** $(\mathbf{D}_0^{(r)}, \mathbf{D}_1^{(r)})$;

$$
\begin{aligned}
E_{(\mathbf{D}_0,\mathbf{D}_1),\mathcal{T}}[Z_i] &= \sum_{k=1}^{m} \frac{\mathbf{F}^{(k)}_{(\mathbf{D}_0,\mathbf{D}_1),t_k}(i,i)}{\boldsymbol{\pi}_s \mathbf{b}^{(1)}_{(\mathbf{D}_0,\mathbf{D}_1),t_1}}, \\
E_{(\mathbf{D}_0,\mathbf{D}_1),\mathcal{T}}[N_{ij}] &= \sum_{k=1}^{m} \frac{\mathbf{D}_0(i,j)\mathbf{F}^{(k)}_{(\mathbf{D}_0,\mathbf{D}_1),t_k}(i,j)}{\boldsymbol{\pi}_s \mathbf{b}^{(1)}_{(\mathbf{D}_0,\mathbf{D}_1),t_1}}, \\
E_{(\mathbf{D}_0,\mathbf{D}_1),\mathcal{T}}[M_{ij}] &= \sum_{k=1}^{m-1} \frac{\mathbf{f}^{(k)}_{(\mathbf{D}_0,\mathbf{D}_1),t_k}(i)\mathbf{D}_1(i,j)\mathbf{b}^{(k+1)}_{(\mathbf{D}_0,\mathbf{D}_1),t_{k+1}}}{\boldsymbol{\pi}_s \mathbf{b}^{(1)}_{(\mathbf{D}_0,\mathbf{D}_1),t_1}} + \frac{\mathbf{f}^{(m)}_{(\mathbf{D}_0,\mathbf{D}_1),t_m}(i)\mathbf{D}_1(i,j)\mathbb{1}}{\boldsymbol{\pi}_s \mathbf{b}^{(1)}_{(\mathbf{D}_0,\mathbf{D}_1),t_1}}
\end{aligned}
\tag{5.13}
$$

With these expectations the M-step equals

$$
\begin{aligned}
\hat{\mathbf{D}}_0(i,j) &= \frac{E_{(\mathbf{D}_0,\mathbf{D}_1),\mathcal{T}}[N_{ij}]}{E_{(\mathbf{D}_0,\mathbf{D}_1),\mathcal{T}}[Z_i]} \quad \text{for } i \neq j, \\
\hat{\mathbf{D}}_1(i,j) &= \frac{E_{(\mathbf{D}_0,\mathbf{D}_1),\mathcal{T}}[M_{ij}]}{E_{(\mathbf{D}_0,\mathbf{D}_1),\mathcal{T}}[Z_i]}, \\
\hat{\mathbf{D}}_0(i,i) &= -\left(\sum_{j=1, i \neq j}^{n} \hat{\mathbf{D}}_0(i,j) + \sum_{j=1}^{n} \hat{\mathbf{D}}_1(i,j) \right).
\end{aligned}
\tag{5.14}
$$

Algorithm 4 is the EM algorithm for MAPs. Although the algorithm looks similar to the EM algorithm for PHDs, there are some significant differences which require additional effort. The vectors $\mathbf{ff}^{(i)}_{(\mathbf{D}_0^{(r)}, \mathbf{D}_1^{(r)}), t_i}$ all have to be precomputed and stored in the loop from line 4 through 6 in the algorithm. For longer traces a huge number of vectors has to be stored and the entries in the vectors may become very small or huge. The latter implies that for a stable implementation vectors have to be rescaled from time to time. The vectors $\mathbf{bb}^{(i)}_{(\mathbf{D}_0^{(r)}, \mathbf{D}_1^{(r)}), t_i}$ that are computed in the backward phase

Table 5.4 Measures and likelihood values of the original MAP, the fitted MAP and the trace

	Original MAP	Trace	Fitted MAP
$E[X^1]$	0.4737	0.4923	0.4923
$E[X^2]$	0.7809	0.7813	0.8201
$E[X^3]$	2.2540	2.0630	2.3480
ρ_1	−0.132	−0.142	−0.158
ρ_2	0.076	0.119	0.097
ρ_3	−0.034	−0.061	−0.055
$\log \mathcal{L}((\mathbf{D}_0,\mathbf{D}_1)\vert\mathcal{T})$	−96.18	–	−92.86

need not be stored but have the same stability problems as the forward vectors such that they also have to be rescaled. It is not possible to use trace aggregation during MAP parameter fitting because the sequence of inter-event times is important and is no longer available if traces are aggregated. This has the consequence that EM algorithms for MAPs can be applied to complete traces with a few thousand entries but not for traces with more than a million entries as they are available in the Internet traffic archive [84]. A general feature of the EM algorithm is that zero elements in a matrix remain zero which means that specific matrix structures, like acyclic matrices $\mathbf{D}_0$, can be used for initialization and the fitted MAP resulting from this initialization has an acyclic matrix $\mathbf{D}_0$.

Example 5.2. We generate a trace with 1000 entries from the following MAP.

$$\mathbf{D}_0 = \begin{bmatrix} -10 & 1 & 1 \\ 1 & -5 & 1 \\ 0 & 0 & -1 \end{bmatrix},\ \mathbf{D}_1 = \begin{bmatrix} 2 & 2 & 4 \\ 0 & 3 & 0 \\ 1 & 0 & 0 \end{bmatrix}.$$

The trace is used as input for Algorithm 4 which generates the following MAP from the trace in about 5 seconds on a standard PC.

$$\mathbf{D}_0 = \begin{bmatrix} -1.02 & 0 & 0 \\ 1.71 & -4.82 & 0 \\ 1.05 & 0 & -8.96 \end{bmatrix},\ \mathbf{D}_1 = \begin{bmatrix} 0 & 1.02 & 0 \\ 3.11 & 0 & 0 \\ 3.30 & 4.08 & 0.53 \end{bmatrix}.$$

Both MAPs differ which has two reasons. First, the representation of a MAP is non-unique and, second, the trace does, of course, not characterize the MAP from which it is generated. Table 5.4 compares some measures, moments and coefficients of autocorrelation, for the original MAP, the fitted MAP and the trace. It can be seen that the fitted MAP has a slightly larger likelihood for the trace than the MAP from which the trace has been generated. Nevertheless, both MAPs show a similar behavior.

The observations from the small example can be generalized. If traces are generated from MAPs with a moderate number of states, then the EM algorithm generates usually a MAP with a similar behavior. However, for real traces or traces from MAPs with a large number of states, the EM algorithm usually requires more effort and yields less good results.

5.3 Two Phase Approaches

An adequate description of traffic processes incorporating correlation requires large amounts of data in a trace. This usually entails an increasing computational complexity and numerical instability when MAP fitting should be directly performed on the complete data trace. For this reason, MAPs can be fitted to a traffic process given by the empirical density or distribution function of the inter-arrival time distribution and to the lag correlation function of the trace.

In the two-phase MAP fitting approach, first an order n PHD with representation $(\boldsymbol{\pi}, \mathbf{D}_0)$ is fitted. We refer to the Sect. 3 where various PHD fitting methods for different inputs like trace data, pdf, cdf or a given number of moments are described. In the second phase, the matrix $\mathbf{D}_1$ is constructed, such that the inter-arrival time distribution of the resulting MAP remains unchanged, i.e, the resulting MAP $(\mathbf{D}_0, \mathbf{D}_1)$ has the stationary distribution $\boldsymbol{\pi}$, and its lag correlation function approximates the correlation of the arrival intervals of the trace.

The role of the resulting PHD representation is not negligible in the two-phase fitting approaches, and, in general, the PHD representation is non-unique [47]. The entries in $(\boldsymbol{\pi}, \mathbf{D}_0)$ have large influence when fitting matrix $\mathbf{D}_1$, since the entries put constraints on the possible values of entries of $\mathbf{D}_1$, thereby limiting the possible range of achievable autocorrelation and the range of joint moments that can be fitted. For a unique representation A. Cumani introduced a canonical form of a PHD (cf. Sect. 2.3.3) that is often used for PHD fitting [18, 134] which is not suitable in the second phase of two-phase MAP fitting approaches since it has only one exit state and does not allow for any flexibility when fitting $\mathbf{D}_1$. The same holds for representations that have only one entry state. Thus, existing transformations aim at increasing the number of entry and exit states [32, 33, 40, 82, 118]. For APHDs with an arbitrary number of states it has been shown in [40] that the hyper-exponential representation results in the maximal first order joint moment for a subsequent MAP fitting step.

Fitting procedures according to the first joint moments and autocorrelation coefficients are presented in Sects. 5.3.1 and 5.3.2.

5.3.1 Joint Moment Fitting

In the construction of the matrix $\mathbf{D}_1$, the following conditions have to be satisfied to ensure that the inter-arrival time distribution determined in the first step remains unchanged, namely $\mathbf{D}_1(i,j) \geq 0$, $\mathbf{D}_1 \mathbb{1} = -\mathbf{D}_0 \mathbb{1}$, and $\boldsymbol{\pi}(-\mathbf{D}_0)^{-1}\mathbf{D}_1 = \boldsymbol{\pi}$ [33,82]. These constraints can be formulated as a linear system of equations. Consider the column vector $\mathbf{x}$ of size n^2, which is composed by the columns of the matrix $\mathbf{D}_1$ as shown below:

$$\mathbf{D}_1 = \left[\, \{\mathbf{D}_1\}_1 \; \{\mathbf{D}_1\}_2 \; \ldots \; \{\mathbf{D}_1\}_n \,\right] \text{ with } \mathbf{x} = \begin{bmatrix} \{\mathbf{D}_1\}_1 \\ \{\mathbf{D}_1\}_2 \\ \vdots \\ \{\mathbf{D}_1\}_n \end{bmatrix}$$

where $\{\mathbf{D}_1\}_i$ denotes the ith column of $\mathbf{D}_1$. The elements of the solution vector $\mathbf{x}$ are non-negative by the first condition such that the resulting matrix $\mathbf{D}_1$ describes a valid MAP. In the coefficient matrix $\mathcal{A}$ and in the column vector $\mathbf{b}$ the necessary conditions are encoded:

$$\underbrace{\begin{bmatrix} \left[\boldsymbol{I}_{n\times n}\right] & \left[\boldsymbol{I}_{n\times n}\right] & \cdots & \left[\boldsymbol{I}_{n\times n}\right] \\ \boxed{\psi} & & & \\ & \boxed{\psi} & & \\ & & \ddots & \\ & & & \boxed{\psi} \end{bmatrix}}_{\mathcal{A}_{2n\times n^2}} \cdot \underbrace{\begin{bmatrix} \\ \mathbf{x} \\ \\ \end{bmatrix}}_{\mathbf{x}_{n^2}} = \underbrace{\begin{bmatrix} \mathbf{d} \\ \\ \boldsymbol{\pi} \end{bmatrix}}_{\mathbf{b}_{2n}}, \tag{5.15}$$

where $\mathbf{d} = -\mathbf{D}_0\mathbb{1}$ and $\psi = \boldsymbol{\pi}(-\mathbf{D}_0)^{-1}$ such that the first n rows of $\mathcal{A}$ correspond to the requirement $\mathbf{D}_1\mathbb{1} = -\mathbf{D}_0\mathbb{1}$, and the remaining rows correspond to the requirement $\boldsymbol{\pi}(-\mathbf{D}_0)^{-1}\mathbf{D}_1 = \boldsymbol{\pi}$. All possible $\mathbf{x}$ vectors, i.e. the matrices $\mathbf{D}_1$ of valid MAPs, satisfying necessary constraints are the solutions of the following system of linear equations and inequalities

$$\mathcal{A}\mathbf{x} = \mathbf{b}, \qquad \mathbf{x} \geq 0. \tag{5.16}$$

For $n = 2$ the matrix $\mathbf{D}_0$ and the vector $\boldsymbol{\pi}$ completely determine the elements of $\mathbf{D}_1$. For $n > 2$ the equation $\mathcal{A}\mathbf{x} = \mathbf{b}$ is under-determined since we have $2n$ equations and n^2 unknowns. In this case, e.g. the simplex algorithm can be used to find a $\mathbf{D}_1$ matrix.

If some joint moments $\hat{\mu}_{ij}$ from the set of measured joint moments $\mathcal{J}$ should be matched by the $\mathbf{D}_1$, the fitting problem can be written as the minimization problem:

$$\min_{\mathbf{D}_1(i,j)\geq 0, \mathbf{D}_1\mathbf{1} = -\mathbf{D}_0\mathbf{1}, \boldsymbol{\pi}(-\mathbf{D}_0)^{-1}\mathbf{D}_1 = \boldsymbol{\pi}} \left(\sum_{\hat{\mu}_{ij}\in\mathcal{J}} \left(\beta_{ij}\frac{\mu_{ij}}{\hat{\mu}_{ij}} - \beta_{ij} \right)^2 \right), \tag{5.17}$$

where $(\boldsymbol{\pi}, \mathbf{D}_0)$ is a valid PHD that is expanded to a MAP representation $(\mathbf{D}_0, \mathbf{D}_1)$ with i,jth order joint moments μ_{ij}. $\mathcal{J}$ is the set of joint moments to be approximated and β_{ij} is a non-negative weight, which allows one to give different weights to the joint

Table 5.5 Measures of joint moments of the traces and of the fitted MAPs

Joint Moment	trace *pAug*	*pAug* MAP(4)	trace *lbl*	*lbl* MAP(3)
μ_{11}	1.6448	1.5243	1.3014	1.294
μ_{22}	106.5773	102.8148	17.4204	17.6220
μ_{33}	54774.5443	51514.3226	890.5772	888.5744
μ_{44}	71016669.2295	69329882.7511	107457.3997	105616.0177
μ_{55}	129600039759.5157	164353821240.3614	21412936.5984	22216642.0776

moments, e.g. to privilege the lower order moments. If the minimum in Eq. (5.17) becomes zero, then the exact solution has been found, otherwise an in terms of the Euclidean norm optimal approximation has been obtained.

The joint moments can be derived from the moment matrix $\mathbf{M} = -\mathbf{D}_0^{-1}$ such that we obtain for the i, jth order joint moment of two consecutive events

$$\mu_{ij} = E[X_1^i, X_2^j] = i!j!\boldsymbol{\pi}\mathbf{M}^{i+1}\mathbf{D}_1\mathbf{M}^j\mathbb{1}, \tag{5.18}$$

and let $\mathbf{x}^i = \boldsymbol{\pi}\mathbf{M}^{i+1}$ and $\mathbf{y}^j = \mathbf{M}^j\mathbb{1}$. In general, the i, jth order joint moment can be expressed as a linear constraint

$$\mu_{ij} = \sum_{r=1}^{n}\sum_{s=1}^{n} \mathbf{x}^i(r)\mathbf{D}_1(r,s)\mathbf{y}^j(s), \tag{5.19}$$

such that the expression for μ_{ij} in Eq. (5.19) can be plugged in for μ_{ij} in the general minimization problem given in Eq. (5.17). E.g., consider the first order joint moment given by $\mu_{11} = E[X_t, X_{t+1}] = \boldsymbol{\pi}(-\mathbf{D}_0)^{-1}\mathbf{P}_s(-\mathbf{D}_0)^{-1}\mathbb{1}$. Then $\mathbf{x}^1 = \boldsymbol{\pi}(-\mathbf{D}_0)^{-1}(-\mathbf{D}_0)^{-1}$, $\mathbf{y}^1 = (-\mathbf{D}_0)^{-1}\mathbb{1}$, and the linear condition on the first order joint moment can be concatenated to the matrix $\mathcal{A}$ and vector $\mathbf{b}$ as shown in Eq. (5.20) where the weight β_{11} is set to one.

$$\underbrace{\begin{bmatrix} [\boldsymbol{I}_{n\times n}] & [\boldsymbol{I}_{n\times n}] & \cdots & [\boldsymbol{I}_{n\times n}] \\ \boxed{\psi} & & & \\ & \boxed{\psi} & & \\ & & \ddots & \\ & & & \boxed{\psi} \\ \boxed{\mathbf{x}^1(1)\cdot\mathbf{y}^1(1)} & \boxed{\mathbf{x}^1(2)\cdot\mathbf{y}^1(1)} & \cdots & \boxed{\mathbf{x}^1(n)\cdot\mathbf{y}^1(n)} \end{bmatrix}}_{\mathcal{A}_{(2n+1)\times n^2}} \cdot \underbrace{\begin{bmatrix} \\ \mathbf{x} \\ \\ \end{bmatrix}}_{\mathbf{x}_{n^2}} = \underbrace{\begin{bmatrix} \mathbf{d} \\ \boldsymbol{\pi} \\ \hat{\mu}_{11} \end{bmatrix}}_{\mathbf{b}_{2n+1}}. \tag{5.20}$$

The resulting problem is a non-negative least squares problem with n^2 variables and $2n$ linear constraints which can be solved with standard algorithms for non-negative least squares problems [107].

Example 5.3. We consider two traces which have been introduced in Example 5.1. Table 5.5 contains the first five joint moments for the traces *lbl*, *pAug*, and the fitted MAPs. For the trace *pAug* the first five joint moments can be approximated by a MAP with 4 states which is visualized below.

$$\mathbf{D}_0 = \begin{bmatrix} -0.1001 & 0.0003 & 0.00043 & 0.00046 \\ 0 & -0.2747 & 0.00117 & 0.1277 \\ 0 & 0 & -2.5629 & 2.5629 \\ 0 & 0 & 0 & -3.4924 \end{bmatrix}, \mathbf{D}_1 = \begin{bmatrix} 0.01789 & 0.0368 & 0.0442 & 0.000001 \\ 0.005 & 0.14079 & 0 & 0 \\ 0 & 0 & 6.82954e-05 & 0 \\ 0.01336 & 0.14421 & 3.33480 & 0 \end{bmatrix}.$$

For the trace *lbl* better approximation of the joint moments can be achieved by a MAP with only 3 states. The following *lbl* MAP(3) was fitted from the *lbl*-trace.

$$\mathbf{D}_0 = \begin{bmatrix} -0.33104 & 0.00099 & 0.00113 \\ 0 & -0.66071 & 0.00236 \\ 0 & 0 & -2.32231 \end{bmatrix}, \mathbf{D}_1 = \begin{bmatrix} 0.15406 & 0.13962 & 0.03524 \\ 0.02358 & 0.51126 & 0.12351 \\ 0.05016 & 0.29920 & 1.97295 \end{bmatrix}.$$

For the trace *pAug* the fitted MAP has a lag 1 coefficient of autocorrelation of 0.16248 which is below 0.200, the value of the trace. The fitted *lbl* MAP has a lag 1 coefficient of autocorrelation of 0.153 which is a very good approximation of 0.155, the value of the *lbl*-trace.

5.3.2 Autocorrelation Fitting

As we have seen in Sect. 5.3.1, for a joint moment fitting, non-negative least squares problems have to be solved. In this section, we will show that autocorrelation fitting represents a linear constrained non-linear optimization problem. In the second fitting step, we start as usual with a PHD $(\boldsymbol{\pi}, \mathbf{D}_0)$ that is expanded to a MAP representation $(\mathbf{D}_0, \mathbf{D}_1)$ to match additionally lag k autocorrelation coefficients $\tilde{\rho}_k$, $k = 1, \ldots, K$ of some observed process.

We first refer to an exact lag 1 correlation fitting problem which has been introduced in [82]. The autocorrelation coefficient of lag 1 between the two consecutive events is defined in Eq. (4.5) (cf. Sect. 4.1.2) and can be transformed with $\lambda = \frac{1}{\boldsymbol{\pi}(-\mathbf{D}_0)^{-1}\mathbb{1}}$ into

$$\begin{aligned} \rho_1 &= \frac{\boldsymbol{\pi}(-\mathbf{D}_0)^{-2}\mathbf{D}_1(-\mathbf{D}_0)^{-1}\mathbb{1} - \left(\boldsymbol{\pi}(-\mathbf{D}_0)^{-1}\mathbb{1}\right)^2}{2\boldsymbol{\pi}(-\mathbf{D}_0)^{-2}\mathbb{1} - \left(\boldsymbol{\pi}(-\mathbf{D}_0)^{-1}\mathbb{1}\right)^2} \\ &= \frac{\lambda^2\left[\boldsymbol{\pi}(-\mathbf{D}_0)^{-2}\mathbf{D}_1(-\mathbf{D}_0)^{-1}\mathbb{1} - \left(\boldsymbol{\pi}(-\mathbf{D}_0)^{-1}\mathbb{1}\right)^2\right]}{\lambda^2\left[2\boldsymbol{\pi}(-\mathbf{D}_0)^{-2}\mathbb{1} - \left(\boldsymbol{\pi}(-\mathbf{D}_0)^{-1}\mathbb{1}\right)^2\right]} \\ &= \frac{\lambda^2\boldsymbol{\pi}(-\mathbf{D}_0)^{-2}\mathbf{D}_1(-\mathbf{D}_0)^{-1}\mathbb{1} - 1}{\lambda^2 2\boldsymbol{\pi}(-\mathbf{D}_0)^{-2}\mathbb{1} - 1}. \end{aligned}$$

Now consider $\mathbf{m} = (-\mathbf{D}_0)^{-1}\mathbb{1}$ such that

$$\lambda^2\boldsymbol{\pi}(-\mathbf{D}_0)^{-2}\mathbf{D}_1\mathbf{m} = \rho_1\left[\lambda^2 2\boldsymbol{\pi}(-\mathbf{D}_0)^{-2}\mathbb{1} - 1\right] + 1,$$

which can be concatenated to the matrix $\mathcal{A}$ and vector $\mathbf{b}$ as shown in Eq. (5.21) with $\boldsymbol{\phi} = \lambda^2\boldsymbol{\pi}(-\mathbf{D}_0)^{-2}$, $\mathbf{m}(i)$ is the ith element of the vector $\mathbf{m}$, and

$$\omega = \tilde{\rho}_1\left[\lambda^2 2\boldsymbol{\pi}(-\mathbf{D}_0)^{-2}\mathbb{1} - 1\right] + 1$$

where $\tilde{\rho}_1$ is the autocorrelation of lag 1 to be approximated by the expanded MAP $(\mathbf{D}_0, \mathbf{D}_1)$ and, e.g. can be estimated from the trace. Eq. (5.21) is similar to Eq. (5.16) mentioned in Sect. 5.3.1.

$$\underbrace{\begin{bmatrix} \begin{bmatrix} I_{n\times n} \end{bmatrix} & \begin{bmatrix} I_{n\times n} \end{bmatrix} & \cdots & \begin{bmatrix} I_{n\times n} \end{bmatrix} \\ \boxed{\psi} & & & \\ & \boxed{\psi} & & \\ & & \ddots & \\ & & & \boxed{\psi} \\ \boxed{\mathbf{m}(1)\boldsymbol{\phi}} & \boxed{\mathbf{m}(2)\boldsymbol{\phi}} & \cdots & \boxed{\mathbf{m}(n)\boldsymbol{\phi}} \end{bmatrix}}_{\mathcal{A}_{(2n+1)\times n^2}} \cdot \underbrace{\begin{bmatrix} \\ \mathbf{x} \\ \\ \end{bmatrix}}_{\mathbf{x}_{n^2}} = \underbrace{\begin{bmatrix} \mathbf{d} \\ \boldsymbol{\pi} \\ \omega \end{bmatrix}}_{\mathbf{b}_{2n+1}}. \tag{5.21}$$

However, if more lag k autocorrelation values should be matched, the $\mathbf{D}_1$ fitting problem has to be formulated as a non-linear minimization problem with $2n$ linear constraints given in Eq. (5.15) and goal function

$$\min_{\mathbf{D}_1(i,j)\geq 0, \mathbf{D}_1\mathbf{1} = -\mathbf{D}_0\mathbf{1}, \boldsymbol{\pi}(-\mathbf{D}_0)^{-1}\mathbf{D}_1 = \boldsymbol{\pi}} \left(\sum_{k=2}^{K}\beta_k(\rho_k - \tilde{\rho}_k)^2\right), \tag{5.22}$$

where ρ_k is lag k autocorrelation coefficient of the fitted MAP with representation $(\mathbf{D}_0, \mathbf{D}_1)$, lag K is the largest lag autocorrelation coefficient that should be considered in the objective function, and weight β_k again may be used to privilege lower lag autocorrelations. Eq. (5.22) is a linearly constrained non-linear optimization problem with the objective function given by the squared difference between the lag k autocorrelations of the observed process and the fitted MAP (cf. Eq. (5.17) in Sect. 5.3.1).

Note, that lag 1 autocorrelation can be expressed as a linear constraint. In contrast, the higher lag autocorrelations result in non-linear constraints, e.g. the lag 2 first order joint moment $E[X_t, X_{t+2}] = \boldsymbol{\pi}(-\mathbf{D}_0)^{-1}\mathbf{P}_s^2(-\mathbf{D}_0)^{-1}\mathbb{1}$ would lead to a term containing squared elements of the matrix $\mathbf{D}_1$.

Example 5.4. We consider again two common traces introduced in Example 5.1. The fitting results are shown below.

$$\text{lbl-trace } \mathbf{D}_0 = \begin{bmatrix} -0.32845 & 0 & 0 & 0 & 0 & 0 \\ 0 & -0.677 & 0.26063 & 0 & 0 & 0 \\ 0 & 0 & -1.01592 & 1.01592 & 0 & 0 \\ 0 & 0 & 0 & -1.9148 & 1.9148 & 0 \\ 0 & 0 & 0 & 0 & -6.01439 & 0 \\ 0 & 0 & 0 & 0 & 0 & -6.0144 \end{bmatrix},$$

$$\mathbf{D}_1 = \begin{bmatrix} 0.04157 & 0.23594 & 00.04623 & 0 & 0.00471 & \\ 0.12054 & 0.26348 & 0 & 0.03112 & 0 & 0.00123 \\ 0 & 0 & 0 & 0 & 0 & 0 \\ 0 & 0 & 0 & 0 & 0 & 0 \\ 0.02603 & 0.88306 & 0 & 5.10410 & 0 & 0.00120 \\ 0.01334 & 0.02466 & 0 & 8.39267e-07 & 0 & 5.97640 \end{bmatrix}$$

$$\text{tudo-trace } \mathbf{D}_0 = \begin{bmatrix} -0.00825 & 0.00508 & 7.6421e-05 & 0.00099 \\ 0 & -0.00942 & 2.71836e-05 & 4.28416e-05 \\ 0 & 0 & -0.02740 & 0.00018 \\ 0 & 0 & 0 & -3.21467 \end{bmatrix},$$

$$\mathbf{D}_1 = \begin{bmatrix} 0.00111 & 0.00092 & 2.5177e-08 & 7.91196e-05 \\ 0.00575 & 2.48645e-10 & 0.00024 & 0.00336 \\ 5.09695e-05 & 9.12383e-05 & 0.02690 & 0.00018 \\ 1.35424e-05 & 0.00020 & 0.00101 & 3.21345 \end{bmatrix}$$

Figure 5.2 shows the course of the lag k autocorrelation coefficient ρ_k for the trace *lbl* and *tudo* and the fitted MAPs. It can be seen that ρ_1 is matched exactly, and at the same time the higher lags are also adequately captured by the fitted MAPs.

5.3.3 Iterative EM Approaches

The computationally most expensive part of the EM algorithm as presented in Sect. 5.2 is the computation of the matrix exponential which has to be done for every trace element in every iteration of the E-step. If we use an EM algorithm as part of a two-step fitting approach and we assume that $\mathbf{D}_0$ and $\boldsymbol{\pi}$ are given from a preceding PHD fitting we can also introduce an EM algorithm that only determines the elements in matrix $\mathbf{D}_1$. Observe from Eqs. (5.11) and (5.12) that the matrix exponential $e^{\mathbf{D}_0 t}$ only has to be evaluated once for every trace element if $\mathbf{D}_0$ is not changed in the iterations of the EM step. In the E-step of the algorithm from Eq. (5.13) we only compute $E_{(\mathbf{D}_0,\mathbf{D}_1),\mathcal{T}}(Z_i)$ and $E_{(\mathbf{D}_0,\mathbf{D}_1),\mathcal{T}}(M_{ij})$ and the M-step of the algorithm in Eq. (5.14) only consists of the computation of a new estimate $\hat{\mathbf{D}}_1(i,j)$. While this approach is very efficient compared to the original EM algorithm, the drawback is that the diagonal elements of $\mathbf{D}_0$ are not adjusted. Furthermore, the $\mathbf{D}_0, \mathbf{D}_1$ pair will not preserve the steady-state distribution $\boldsymbol{\pi}$. To preserve the distribution given by $\mathbf{D}_0$ and $\boldsymbol{\pi}$ the $\mathbf{D}_1$ matrix has to be repaired after a few iterations of the EM algorithm. This can be done by formulating an optimization problem

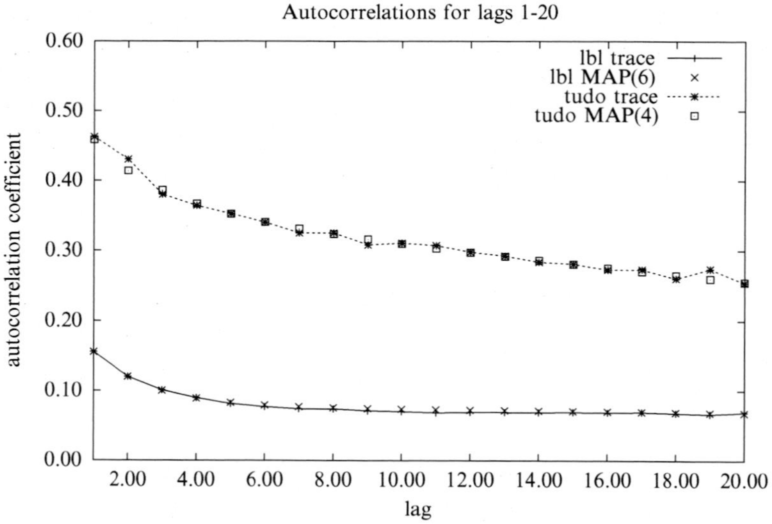

Fig. 5.2 Lag k coefficient of autocorrelation for the traces *lbl* and *tudo*

with the constraints $-\mathbf{D}_0\mathbb{1} = \mathbf{D}'_1\mathbb{1}$ and $\boldsymbol{\pi}(-\mathbf{D}_0)^{-1}\mathbf{D}'_1 = \boldsymbol{\pi}$ minimizing the entry-wise Frobenius norm $\|\mathbf{D}'_1 - \hat{\mathbf{D}}_1\|^2$ where $\hat{\mathbf{D}}_1$ is the matrix from the EM algorithm and $\mathbf{D}'_1$ is a matrix that preserves the distribution and has a minimal difference to $\hat{\mathbf{D}}_1$. Then the EM algorithm can be continued with matrix $\mathbf{D}'_1$. In summary, the approach iterates between a few EM steps that find a matrix $\hat{\mathbf{D}}_1$ with a large likelihood and a repair step that will probably slightly decrease the likelihood again, but results in a matrix that respects the distribution that was given.

5.4 Fitting of the Counting Process

In Sect. 3.1.2 the general EM algorithm for PHDs was extended to the case where the number of events in finite intervals are available and not the detailed inter-event times. Similar extensions are possible for MAPs and have been proposed in [26, 133].

Consider again a grouped trace $\tilde{\mathcal{T}}$ as defined in Sect. 3.1.1 that contains m_k events in the kth interval $(T_{k-1}, T_k], k = 1,..,M$ with interval length $\Delta_k = T_k - T_{k-1}$ and a MAP $(\mathbf{D}_0, \mathbf{D}_1)$ of order n for which the entries of the two matrices should be fitted such that the likelihood of generating the observed number of events in the intervals is maximized.

Again we use Z_i, N_{ij} and M_{ij} as the total time spent in state i, the total observed number of jumps from state i to state j $(i \neq j)$ without generating an event and the total observed number of jumps from state i to j $(i, j \in \mathcal{S})$ when generating an event, respectively.

The interpretation for a MAP as counting process was given in Sect. 4.1.4 and will be picked up in the following. The EM algorithm for the counting process uses forward and backward likelihood vectors similar to the EM algorithm for the inter-event times. However, since we are fitting a counting process, the vectors are defined as block vectors as in Sect. 4.1.4. Let $\mathbf{Q}^c_{m_k}$ be the upper left $(n \cdot m_k + n) \times (n \cdot m_k + n)$ submatrix of $\mathbf{Q}^c$ from Eq. (4.7) and define

$$\tilde{\mathbf{D}_1} = [\mathbf{0} \quad \cdots \quad \mathbf{0} \quad \mathbf{I}]^T [\mathbf{I} \quad \mathbf{0} \quad \cdots \quad \mathbf{0}].$$

Moreover, let $\tilde{\boldsymbol{\pi}} = [\boldsymbol{\pi}, \mathbf{0}, \cdots, \mathbf{0}]$ and $\tilde{\mathbb{1}} = [\mathbf{0}, \cdots, \mathbf{0}, \mathbf{1}]^T$. Using these notations we can define the forward and backward likelihood vectors as follows. Let $\tilde{\mathbf{ff}}^{(k)}_{(\mathbf{D}_0,\mathbf{D}_1)}$ and $\tilde{\mathbf{bb}}^{(k)}_{(\mathbf{D}_0,\mathbf{D}_1)}$ be the forward and backward block vector before and after the kth interval, i.e.

$$\begin{aligned} \tilde{\mathbf{ff}}^{(k)}_{(\mathbf{D}_0,\mathbf{D}_1)} &= \begin{cases} \tilde{\boldsymbol{\pi}} & \text{if } k = 1 \\ \tilde{\mathbf{ff}}^{(k-1)}_{(\mathbf{D}_0,\mathbf{D}_1)} e^{\mathbf{Q}^c_{m_{k-1}} \cdot \Delta_{k-1}} \tilde{\mathbf{D}_1} & \text{if } 1 < k \leq M \end{cases} \\ \tilde{\mathbf{bb}}^{(k)}_{(\mathbf{D}_0,\mathbf{D}_1)} &= \begin{cases} \tilde{\mathbf{D}_1} \tilde{\mathbb{1}} & \text{if } k = M \\ \tilde{\mathbf{D}_1} e^{\mathbf{Q}^c_{m_{k+1}} \cdot \Delta_{k+1}} \tilde{\mathbf{bb}}^{(k+1)}_{(\mathbf{D}_0,\mathbf{D}_1)} & \text{if } 1 \leq i < M \end{cases} \end{aligned} \tag{5.23}$$

Forward and backward block vectors for the kth interval are then given by

$$\begin{aligned} \tilde{\mathbf{f}}^{(k)}_{(\mathbf{D}_0,\mathbf{D}_1),t} &= \tilde{\mathbf{ff}}^{(k)}_{(\mathbf{D}_0,\mathbf{D}_1)} e^{\mathbf{Q}^c_{m_k} \cdot t} \\ \tilde{\mathbf{b}}^{(k)}_{(\mathbf{D}_0,\mathbf{D}_1),t} &= e^{\mathbf{Q}^c_{m_k} \cdot t} \tilde{\mathbf{bb}}^{(k)}_{(\mathbf{D}_0,\mathbf{D}_1)} \end{aligned} \tag{5.24}$$

Vector $\tilde{\mathbf{f}}^{(k)}_{(\mathbf{D}_0,\mathbf{D}_1),t}$ consists of blocks $\mathbf{f}^{(k)}_{(\mathbf{D}_0,\mathbf{D}_1),t,x}$, i.e.

$$\tilde{\mathbf{f}}^{(k)}_{(\mathbf{D}_0,\mathbf{D}_1),t} = \left[\mathbf{f}^{(k)}_{(\mathbf{D}_0,\mathbf{D}_1),t,0} \quad \mathbf{f}^{(k)}_{(\mathbf{D}_0,\mathbf{D}_1),t,1} \quad \cdots \quad \mathbf{f}^{(k)}_{(\mathbf{D}_0,\mathbf{D}_1),t,m_k}\right].$$

Each vector $\mathbf{f}^{(k)}_{(\mathbf{D}_0,\mathbf{D}_1),t,x}$ describes the probability for x arrivals in t time units within the kth interval given the arrivals in the previous intervals. In a similar way $\tilde{\mathbf{b}}^{(k)}_{(\mathbf{D}_0,\mathbf{D}_1),t}$ can be divided into vectors $\mathbf{b}^{(k)}_{(\mathbf{D}_0,\mathbf{D}_1),t,x}$. Using these vectors we can define matrix

$$\tilde{\mathbf{F}}^{(k)}_{(\mathbf{D}_0,\mathbf{D}_1),t,x} = \sum_{l=0}^{x} \int_0^t \left(\mathbf{f}^{(k)}_{(\mathbf{D}_0,\mathbf{D}_1),t-u,l}\right)^T \left(\mathbf{b}^{(k)}_{(\mathbf{D}_0,\mathbf{D}_1),u,x-l}\right)^T du \tag{5.25}$$

and matrix

$$\tilde{\mathbf{G}}^{(k)}_{(\mathbf{D}_0,\mathbf{D}_1),t,x} = \sum_{l=0}^{x-1} \int_0^t \left(\mathbf{f}^{(k)}_{(\mathbf{D}_0,\mathbf{D}_1),t-u,l}\right)^T \left(\mathbf{b}^{(k)}_{(\mathbf{D}_0,\mathbf{D}_1),u,x-l-1}\right)^T du \tag{5.26}$$

These values can be computed using either uniformization or by solving a set of differential equations as given in Eq. (4.8) with e.g. $\psi_k^\tau = \mathbf{f}^{(i)}_{(\mathbf{D}_0,\mathbf{D}_1),\tau,k}$. We now show how the vectors and matrices can be computed using uniformization. α, $\mathbf{P}_0$ and $\mathbf{P}_1$ are defined as in Sect. 4.1.2. Define vectors $\boldsymbol{\phi}^{(l)}_{(\mathbf{D}_0,\mathbf{D}_1),k,i}$ of length n with $1 \le k \le M$ and $0 \le i \le m_k$. For $l = 0$ the vectors are defined as

$$\boldsymbol{\phi}^{(0)}_{(\mathbf{D}_0,\mathbf{D}_1),k,i} = \begin{cases} \boldsymbol{\pi}_s & \text{if } k = 0 \text{ and } i = 0, \\ \mathbf{f}^{(k-1)}_{(\mathbf{D}_0,\mathbf{D}_1),\Delta_{k-1},m_{k-1}} & \text{if } k > 0 \text{ and } i = 0, \\ \mathbf{0} & \text{otherwise,} \end{cases} \tag{5.27}$$

and for $l > 0$

$$\boldsymbol{\phi}^{(l)}_{(\mathbf{D}_0,\mathbf{D}_1),k,i} = \begin{cases} \boldsymbol{\phi}^{(l-1)}_{(\mathbf{D}_0,\mathbf{D}_1),k,i}\mathbf{P}_0 & \text{if } i = 0, \\ \boldsymbol{\phi}^{(l-1)}_{(\mathbf{D}_0,\mathbf{D}_1),k,i}\mathbf{P}_0 + \boldsymbol{\phi}^{(l-1)}_{(\mathbf{D}_0,\mathbf{D}_1),k,i-1}\mathbf{P}_1 & \text{otherwise.} \end{cases} \tag{5.28}$$

Then

$$\mathbf{f}^{(k)}_{(\mathbf{D}_0,\mathbf{D}_1),t,i)} = \sum_{l=0}^{\infty} \beta(\alpha t, l)\boldsymbol{\phi}^{(l)}_{(\mathbf{D}_0,\mathbf{D}_1),k,i}. \tag{5.29}$$

Similarly the equations for the backward vectors can be defined using vectors $\boldsymbol{\varphi}^{(l)}_{(\mathbf{D}_0,\mathbf{D}_1),k,i}$ with

$$\boldsymbol{\varphi}^{(0)}_{(\mathbf{D}_0,\mathbf{D}_1),k,i} = \begin{cases} \mathbb{1} & \text{if } k = M \text{ and } i = 0, \\ \mathbf{b}^{(k+1)}_{(\mathbf{D}_0,\mathbf{D}_1),\Delta_{k-1},m_{k-1}} & \text{if } k < M \text{ and } i = 0, \\ \mathbf{0} & \text{otherwise,} \end{cases} \tag{5.30}$$

and for $l > 0$

$$\boldsymbol{\varphi}^{(l)}_{(\mathbf{D}_0,\mathbf{D}_1),k,i} = \begin{cases} \mathbf{P}_0\boldsymbol{\varphi}^{(l-1)}_{(\mathbf{D}_0,\mathbf{D}_1),k,i} & \text{if } i = 0, \\ \mathbf{P}_0\boldsymbol{\varphi}^{(l-1)}_{(\mathbf{D}_0,\mathbf{D}_1),k,i} + \mathbf{P}_1\boldsymbol{\varphi}^{(l-1)}_{(\mathbf{D}_0,\mathbf{D}_1),k,i-1} & \text{otherwise.} \end{cases} \tag{5.31}$$

Then

$$\mathbf{b}^{(k)}_{(\mathbf{D}_0,\mathbf{D}_1),t,i)} = \sum_{l=0}^{\infty} \beta(\alpha t, l)\boldsymbol{\varphi}^{(l)}_{(\mathbf{D}_0,\mathbf{D}_1),k,i}. \tag{5.32}$$

Matrices $\tilde{\mathbf{F}}^{(k)}_{(\mathbf{D}_0,\mathbf{D}_1),t,x}$ and $\tilde{\mathbf{G}}^{(k)}_{(\mathbf{D}_0,\mathbf{D}_1),t,x}$ are then given by

$$\begin{aligned}\tilde{\mathbf{F}}^{(k)}_{(\mathbf{D}_0,\mathbf{D}_1),t,x} &= \frac{1}{\alpha}\sum_{l=0}^{\infty}\beta(\alpha t,l+1)\sum_{h=0}^{l}\sum_{y=0}^{x}\left(\boldsymbol{\phi}^{(h)}_{(\mathbf{D}_0,\mathbf{D}_1),k,y}\right)^T\left(\boldsymbol{\varphi}^{(l-h)}_{(\mathbf{D}_0,\mathbf{D}_1),k,x-y}\right)^T,\\ \tilde{\mathbf{G}}^{(k)}_{(\mathbf{D}_0,\mathbf{D}_1),t,x} &= \frac{1}{\alpha}\sum_{l=0}^{\infty}\beta(\alpha t,l+1)\sum_{h=0}^{l}\sum_{y=0}^{x-1}\left(\boldsymbol{\phi}^{(h)}_{(\mathbf{D}_0,\mathbf{D}_1),k,y}\right)^T\left(\boldsymbol{\varphi}^{(l-h)}_{(\mathbf{D}_0,\mathbf{D}_1),k,x-1-y}\right)^T,\end{aligned} \tag{5.33}$$

where $0 \le x \le m_k$ for matrix $\tilde{\mathbf{F}}$ and $0 < x \le m_k$ for matrix $\tilde{\mathbf{G}}$.

After the matrices have been computed, we can use them in the E-step to obtain

$$\begin{aligned}E_{(\mathbf{D}_0,\mathbf{D}_1),\mathcal{T}}[Z_i] &= \sum_{k=1}^{M}\sum_{l=0}^{m_k}\tilde{\mathbf{F}}^{(k)}_{(\mathbf{D}_0,\mathbf{D}_1),l,\Delta_k}(i,i),\\ E_{(\mathbf{D}_0,\mathbf{D}_1),\mathcal{T}}[N_{ij}] &= \sum_{k=1}^{M}\sum_{l=0}^{m_k}\mathbf{D}_0(i,j)\tilde{\mathbf{F}}^{(k)}_{(\mathbf{D}_0,\mathbf{D}_1),l,\Delta_k}(i,j),\\ E_{(\mathbf{D}_0,\mathbf{D}_1),\mathcal{T}}[M_{ij}] &= \sum_{k=1}^{M}\sum_{l=1}^{m_k}\mathbf{D}_1(i,j)\tilde{\mathbf{G}}^{(k)}_{(\mathbf{D}_0,\mathbf{D}_1),l,\Delta_k}(i,j).\end{aligned} \tag{5.34}$$

The M-step can be performed using Eq. (5.14). Thus, Algorithm 4 can be applied using Eq. (5.23) in step 4, Eqs. (5.23)–(5.26) in step 8, and Eq. (5.34) in step 10.

5.5 Concluding Remarks

Although a large number of methods is nowadays available to fit the parameters of a MAP, parameter fitting for MAPs is still much more complex than for PHDs. This means that the effort is high and the quality of the resulting MAPs is not always completely satisfactory. In particular, it is not yet clear which method is best suited for a given data set. EM algorithms, which are successfully applied for PHDs, require often a huge effort when used to compute the parameter of a MAP such that they are only usable for small traces and MAPs with a small number of states. The fitting of MAP parameters according to moments or joint moments is much more efficient but as already mentioned for PHD fitting, higher order moments or joint moments result in highly non-linear equations and most times cannot be estimated in a reliable way. Currently two phase approaches that start with PHD fitting and then expand the PHD into a MAP seem to be promising. However, the quality of the second step depends on the representation of the PHD generated in the first step. Since the representation is non-unique, there is a need to find the best representation which is unknown. Nevertheless, even if the parameter fitting of MAPs is still a challenge, the MAPs which can be generated with available approaches are often better approximations of observed behavior than other models to describe correlated observations like multivariate Normal or Lognormal distributions or ARMA processes [105].

Parameter fitting for MMAPs or BMAPs can be done with extended versions of the methods used for MAPs [39, 97] and is faced with similar problems. For RAPs and MRAPs no specific methods for parameter fitting are known yet.

Chapter 6
Stochastic Models Including PH Distributions and MAPs

PHDs and MAPs are used to define inter-event times at various levels and in different model types. Originally, phase-type representations of inter-event times are used in models that are mapped on Markov processes and are solved numerically. However, this is only one application area. Due to their flexibility phase-type distributions may be applied as well in simulation models and serve as a base for approximate solution techniques. It is not intended to give a complete overview of model based analysis using PHDs and MAPs, instead, we outline some classical and more recent examples of models that are based on PHDs and MAPs. As examples we introduce queueing systems, availability models and simulation models.

6.1 Queueing Systems

Queueing systems [108, 152] are the classical model formalism for discrete event systems. PHDs have been used to model service times and MAPs are a common model to describe arrival processes in single queues and also in networks of queues. In this way, a queueing model can be mapped onto a Markov chain which can be analyzed numerically, as long as the state space is not too large. As examples we consider single queues and single class queueing networks with PHDs and MAPs.

6.1.1 Single Queues

The simplest model are single queues with a single class of customers. The model is completely characterized by the arrival process, the service process, the number of servers, the capacity of the waiting room and the scheduling discipline. We consider here single server queues with an unlimited waiting room and FCFS scheduling.

P. Buchholz et al., *Input Modeling with Phase-Type Distributions and Markov Models: Theory and Applications*, SpringerBriefs in Mathematics,
DOI 10.1007/978-3-319-06674-5_6,

This results in queues of the type *PH/PH/1*, *MAP/PH/1* and *MAP/MAP/1* (see [95, 151] for details). In the following the analysis of *MAP/MAP/1* queues is briefly introduced which includes the other two systems as specific cases.

In a *MAP/MAP/1* queue, the arrival and the service process are characterized by MAPs. Let $(\mathbf{D}_0^a, \mathbf{D}_1^a)$ be the MAP of size n^a describing the arrival process and $(\mathbf{D}_0^s, \mathbf{D}_1^s)$ the MAP of size n^s for the service process, respectively. Since the waiting room is unlimited, the system may potentially contain an unlimited number of customers and the state space is unbounded. States can be grouped into levels according to the population in the system. Each level has $n^a n^s$ states that correspond to the states of the arrival and service process, respectively. We assume that the empty system keeps the state of the service process which, however, does not change as long as the system remains empty. The generator matrix of the Markov process underlying the *MAP/MAP/1* queue has the following structure which is denoted as a quasi-birth-death process (QBD) [104, 125].

$$\mathbf{Q} = \begin{bmatrix} \bar{\mathbf{A}}_0 & \mathbf{A}_1 & \mathbf{0} & \cdots & \\ \mathbf{A}_{-1} & \mathbf{A}_0 & \mathbf{A}_1 & & \\ \mathbf{0} & \mathbf{A}_{-1} & \mathbf{A}_0 & \mathbf{A}_1 & \\ \vdots & & \ddots & \ddots & \ddots \end{bmatrix}$$

The structure is repetitive and all submatrices are of size $n^a n^s$. Submatrices can be expressed in terms of the MAP matrices as follows.

$$\bar{\mathbf{A}}_0 = \mathbf{D}_0^a \otimes \mathbf{I}_{n^s},\ \mathbf{A}_0 = \mathbf{D}_0^a \oplus \mathbf{D}_0^s,\ \mathbf{A}_{-1} = \mathbf{I}_{n^a} \otimes \mathbf{D}_1^s \ \text{ and } \mathbf{A}_1 = \mathbf{D}_1^a \otimes \mathbf{I}_{n^s}$$

where $\mathbf{I}_n$ is the identity matrix of order n.

The stationary distribution of the system is the solution of the system $\boldsymbol{\psi}\mathbf{Q} = \mathbf{0}$ and $\boldsymbol{\psi}\mathbb{1} = 1$ which can be solved for QBDs even if the system of equations contains an infinite number of variables. The stationary vector is decomposed into subvector $\boldsymbol{\psi}_i$ $(i = 0, 1, \ldots)$ of length $n^a n^s$ where $\boldsymbol{\psi}_i$ includes the detailed state probabilities for states with population i in the queue. From vector $\boldsymbol{\psi}$ performance quantities can be derived. E.g., the mean population equals $\sum_{i=1}^{\infty} i \cdot \boldsymbol{\psi}_i \mathbb{1}$, and $\boldsymbol{\psi}_i \mathbb{1}$ is the probability of finding i customers in the queue.

Analysis of the *MAP/MAP/1* system is based on the pioneering work of Neuts on matrix analytic methods [125] and subsequently developed efficient algorithms [16, 104, 145]. We only outline the basic solution approach and refer to the literature for further details.

For the existence of the stationary distribution it is necessary that the system is in equilibrium which means that the mean arrival rate is less than the mean service rate. Let $\boldsymbol{\pi}^a(-\mathbf{D}_0^a)^{-1}\mathbf{D}_1^a = \boldsymbol{\pi}^a$, $\boldsymbol{\pi}^a\mathbb{1} = 1$ and $\boldsymbol{\pi}^s(-\mathbf{D}_0^s)^{-1}\mathbf{D}_1^s = \boldsymbol{\pi}^s$, $\boldsymbol{\pi}^s\mathbb{1} = 1$ be the embedded initial vectors of the MAPs for the arrival and service time. Then $\lambda^a = \left(\boldsymbol{\pi}^a(-\mathbf{D}_0^a)^{-1}\mathbb{1}\right)^{-1}$ and $\lambda^s = \left(\boldsymbol{\pi}^s(-\mathbf{D}_0^s)^{-1}\mathbb{1}\right)^{-1}$ are the mean arrival and service rate and $\rho = \lambda^a/\lambda^s < 1$ is the utilization of the server.

The solution of QBDs relies on the relation

$$\psi_i = \psi_0 \mathbf{R}^i \tag{6.1}$$

for all $i > 0$ where matrix $\mathbf{R}$ is the minimal non-negative solution of the quadratic matrix equation

$$\mathbf{A}_1 + \mathbf{R}\mathbf{A}_0 + \mathbf{R}^2\mathbf{A}_{-1} = \mathbf{0} \tag{6.2}$$

which exists for queues in equilibrium. Matrix $\mathbf{R}$ can be computed using the logarithmic reduction approach [104]. The missing vector ψ_0 is the solution of

$$\psi_0\left(\bar{\mathbf{A}}_0 + \mathbf{R}\mathbf{A}_{-1}\right) = \mathbf{0} \text{ and } \sum_{i=0}^{\infty} \psi_i \mathbb{1} = \psi_0 \sum_{i=0}^{\infty} \mathbf{R}^i \mathbb{1} = \psi_0 (\mathbf{I} - \mathbf{R})^{-1} \mathbb{1} = 1 \tag{6.3}$$

Example 6.1. As an example we analyze a simple MAP/MAP/1 system. Arrivals are generated from a MAP with 3 states and the following matrices.

$$\mathbf{D}_0^a = \begin{bmatrix} -0.3340 & 0.0002 & 0.0002 \\ 0.0000 & -0.6301 & 0.0004 \\ 0.0000 & 0.0000 & -1.9977 \end{bmatrix} \text{ and } \mathbf{D}_1^a = \begin{bmatrix} 0.1637 & 0.1110 & 0.0599 \\ 0.0174 & 0.4776 & 0.1347 \\ 0.0535 & 0.2196 & 1.7246 \end{bmatrix}.$$

Parameters of the MAP have been fitted according to the moments and joint moments of the trace *lbl* from the internet traffic archive [84] where the interarrival times have been scaled such that the mean interarrival time becomes 1. The squared coefficient of variation of the MAP is 1.94 and the autocorrelation coefficients of lag 1 through 3 are 0.146, 0.089 and 0.055. Service times are specified by the following MAP with 2 states.

$$\mathbf{D}_0^s = \begin{bmatrix} -2.00 & 0.00 \\ 0.00 & -0.50 \end{bmatrix} \text{ and } \mathbf{D}_1^s = \begin{bmatrix} 1.95 & 0.05 \\ 0.05 & 0.45 \end{bmatrix}.$$

The mean service time equals 0.8, the squared coefficient of variation is 2.12 and the first three lags of the autocorrelation coefficient are 0.232, 0.203 and 0.177.

The queue with MAP arrival and service and another queue where the MAPs are substituted by PH distributions keeping the distribution of interarrival and service times but neglecting correlation are analyzed. The probabilities of finding 0 through 50 customers in the system in steady state are printed on a logarithmic scale in Fig. 6.1. It can be seen that the neglection of autocorrelations in the PH/PH/1 system results in a different behavior of the system. In particular the tail probabilities are much smaller when correlation is not considered in the model. This might have dramatic consequences if the results are used for capacity planing or dimensioning of real systems. In the simple example, the probability of finding more than 50 customers in the system without correlation is only 0.3%. However, in the *MAP/MAP/1* queue with correlation the probability of finding more than 50

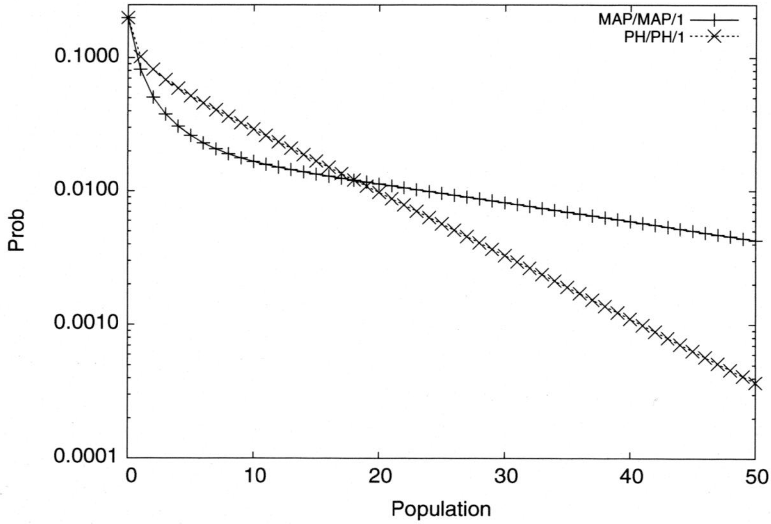

Fig. 6.1 Probability of buffer filling 0 through 50 for the *PH/PH/1* and *MAP/MAP/1* queue

customers equals 12.9%. Consequently, the use of the model without correlation for dimensioning the buffer would result in a buffer size that is much too small to reach the required quality of service.

The basic approach presented in this section can be used in a similar way to analyze finite capacity systems [16], as a building block for the approximate analysis of queueing networks containing queues with MAP or PHD services [50,70,79] and for QBDs where the submatrices depend on the level [49].

6.1.2 *Queueing Networks*

Queueing networks (QNs) are a basic modeling paradigm for many discrete event systems [95,96,108]. We consider here only networks with a single customer class which are either closed, i.e. have a fixed and constant population, or open and all queues are capacity restricted. Service times are specified by PHDs or MAPs and arrivals, if available, are specified by MAPs. This implies that the models describe CTMCs with a finite state space which, at least theoretically, can be analyzed by numerical techniques. However, for large state spaces also simulation or approximate techniques may be applied. The use of PHDs and MAPs allows one to model service times and arrival processes much more detailed than possible with exponential distributions which are traditionally used in QNs [108].

We begin with closed queueing networks that contain N statistically identical customers and K queues or stations. The service time at station k is characterized

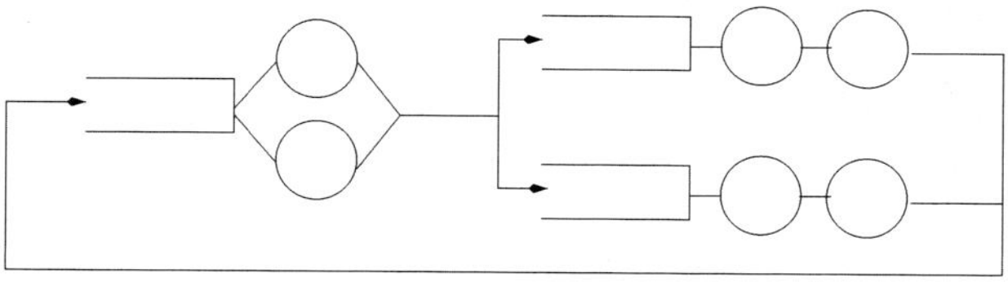

Fig. 6.2 Central server example QN

by PHD $(\boldsymbol{\pi}^k, \mathbf{D}_0^k)$ of size n^k. MAPs may also be used for the specification of service processes but will not be considered in our example. If a customer leaves queue k it enters with probability r_{kl} queue l. We assume that the routing is irreducible such that every queue is reachable from every other queue.

The state space $\mathcal{S}$ of the closed QN can be decomposed into subsets denoted as macro states. Macro states are defined by the distribution of customers among the queues. This implies that the QN has $\binom{N+K-1}{N}$ macro states. In a macro state, the distribution of customers over the queues is fixed and the detailed state results from the state of the PHDs that belong to non-empty queues. If queues $\mathcal{I} \subseteq \{1,\ldots,K\}$ are non-empty, then the macro state contains $\prod_{k\in\mathcal{I}} n^k$ detailed states. This shows that the use of PHDs with a larger number of phases results in huge state spaces, even in relatively small QNs.

The state space and generator matrix $\mathbf{Q}$ of a closed QN with phase-type distributed service times can be naturally structured and this structure can be applied for an efficient solution. We briefly present the basic structure and refer for further details to the literature [28, 30, 49]. Macro states are described by vector $\mathbf{n}$ of length K where $\mathbf{n}(k)$ denotes the number of customers in queue k, if the QN is in a state from macro state $\mathbf{n}$. Of course, $\mathbf{n}(k) \geq 0$ and $\sum_{k=1}^{K} \mathbf{n}(k) = N$ has to hold for all macro states. Macro states can be ordered lexicographically resulting in the macro state space $\tilde{\mathcal{S}}$ and the generator matrix $\mathbf{Q}$ of the Markov chain is block-structured into $|\tilde{\mathcal{S}}| \times |\tilde{\mathcal{S}}|$ blocks $\mathbf{Q}_{\mathbf{n},\mathbf{n}'}$. Submatrix $\mathbf{Q}_{\mathbf{n},\mathbf{n}'}$ is only non-zero, if $\mathbf{n}' = \mathbf{n}$ or $\mathbf{n}'$ results from $\mathbf{n}$ by moving one customer from one queue to another. Formally this means that $\mathbf{n}' = \mathbf{n} + \mathbf{e}_k - \mathbf{e}_l$ where $\mathbf{e}_k$ is a vector with 1 in position k and 0 elsewhere, and $\mathbf{n}(k) > 0$, $r_{kl} > 0$. Each non-zero submatrix can then be generated from the matrices and vectors describing the PHD using Kronecker products and sums as shown in [28, 30, 49]. Different numerical solution techniques are applicable to compute the stationary vector and the result measures for the QN. These methods often exploit the specific matrix structure for a more efficient solution, if the size of the state space is large [30, 49].

Example 6.2. Figure 6.2 shows a simple QN example model which describes a central server system where the central queue has two phase hyper-exponentially distributed service times and the remaining two queues have Erlang 2 service time distributions. The overall number of macro states for the QN with N customers equals $\binom{N+2}{N}$. E.g., for $N = 3$ the vectors $(3,0,0), (2,1,0), (2,0,1), \ldots, (0,1,2), (0,0,3)$

build the macro state space. The number of detailed states in a macro state equals 2 if only one queue has a non-zero population, it equals 4 with two queues with non-zero population and 8 if all three queues have a non-zero population.

It is easy to show that with population $N > 0$, 3 macro states exist with non-zero population in one queue, $3(N-1)$ macro states with non-zero population in two queues and $(N-2)(N-1)/2$ with non-zero population in all three queues. This implies that the number of states of the Markov chain for the example equals

$$6 + 12(N-1) + 4(N-2)(N-1).$$

If the number of queues or the sizes of the PHDs are increased, then the size of the state space growth rapidly.

Apart from closed QNs also open QNs with capacity restricted queues can be modeled as finite Markov chains. In this case each queue has a capacity restriction of N_k $(k = 1,\ldots,K)$. Customers that try to enter a queue which is full either get lost or remain in the previous queue and block the sever. The latter QNs are denoted as blocking QNs [10], the former as loss networks. We briefly introduce loss networks here. Like in closed QNs, routing between queues k and l is described by routing probabilities r_{kl} but an additional index 0 is added to describe customers leaving the system, $\sum_{l=0}^{K} r_{kl} = 1$ has to hold. Service times are again described by PHDs. Additionally, arrivals enter the QN. The arrival process to queue k is characterized by a MAP $\left(\mathbf{D}_0^{(a,k)}, \mathbf{D}_1^{(a,k)}\right)$.

The state space of a QN with losses can also be decomposed into macro states according to the population in different queues. Let $\mathbf{n}$ be again the population vector. In this case vector $\mathbf{n}$ is a valid population vector if $0 \leq \mathbf{n}(k) \leq N_k$ for all $k = 1,\ldots,K$. A transition between states from macro state $\mathbf{n}$ and $\mathbf{n}'$ is possible if $\mathbf{n}' = \mathbf{n} - \mathbf{e}_k$ $(r_{k0} > 0,\ \mathbf{n}(k) > 0)$, $\mathbf{n}' = \mathbf{n} + \mathbf{e}_k$ $(\mathbf{n}(k) < N_k)$ or $\mathbf{n}' = \mathbf{n} - \mathbf{e}_k + \mathbf{e}_l$ $(r_{kl} > 0$, $\mathbf{n}(k) > 0$, $\mathbf{n}(l) < N_l)$. The different submatrices $\mathbf{Q}_{\mathbf{n},\mathbf{n}'}$ can again be generated from the matrices for the PHDs and MAPs describing arrival and service processes. The size of the state space grows rapidly because the number of macro states equals $\prod_{k=1}^{K}(N_k+1)$ and a macro state contains up to $\prod_{k=1}^{K} n^{sk} n^{ak}$ states where n^{ak} equals the number of states of the MAP for the arrivals to queue k and n^{sk} equals the number of states of the PHD describing service at queue k. Consequently, only for small QNs a numerical analysis of the system $\boldsymbol{\psi}\mathbf{Q} = \mathbf{0}$ is manageable. In other cases simulation or approximations have to be applied.

We briefly outline an approximation approach for this class of QNs which has been proposed in [70] and uses fitting approaches for MAPs as building blocks for the approximate analysis of the network. The idea is to decompose the QN into single queues and substitute the environment of the queue by a MAP that mimics arrivals to the queue. In this way analysis of the overall QN is substituted by the repeated analysis of $MAP/PH/1/N$ queues where in the original approach N might be infinite which will be excluded here. The generator matrix of a $MAP/PH/1/N$ equals

$$\mathbf{Q} = \begin{bmatrix} \bar{\mathbf{A}}_0 & \mathbf{A}_1 & \mathbf{0} & \cdots & \cdots & \mathbf{0} \\ \mathbf{A}_{-1} & \mathbf{A}_0 & \mathbf{A}_1 & \ddots & & \vdots \\ \mathbf{0} & \mathbf{A}_{-1} & \mathbf{A}_0 & \mathbf{A}_1 & \ddots & \vdots \\ \vdots & & \ddots & \ddots & \ddots & \mathbf{0} \\ \vdots & & & \mathbf{A}_{-1} & \mathbf{A}_0 & \mathbf{A}_1 \\ \mathbf{0} & \cdots & \cdots & \mathbf{0} & \mathbf{A}_{-1} & \mathbf{A}_0 + \mathbf{A}_1 \end{bmatrix},$$

where matrices $\bar{\mathbf{A}}_0$, $\mathbf{A}_1$, $\mathbf{A}_0$ and $\mathbf{A}_{-1}$ are defined in Sect. 6.1.1. The output process of this system is described by the following MAP.

$$\mathbf{D}_0 = \begin{bmatrix} \bar{\mathbf{A}}_0 & \mathbf{A}_1 & \mathbf{0} & \cdots & \cdots & \mathbf{0} \\ \mathbf{0} & \mathbf{A}_0 & \mathbf{A}_1 & \ddots & & \vdots \\ \vdots & \ddots & \mathbf{A}_0 & \mathbf{A}_1 & \ddots & \\ \vdots & \ddots & \ddots & \ddots & \ddots & \mathbf{0} \\ \vdots & & \ddots & \ddots & \mathbf{A}_0 & \mathbf{A}_1 \\ \mathbf{0} & \cdots & \cdots & \cdots & \mathbf{0} & \mathbf{A}_0 + \mathbf{A}_1 \end{bmatrix} \text{ and } \mathbf{D}_1 = \begin{bmatrix} \mathbf{0} & \cdots & \cdots & \cdots & \mathbf{0} \\ \mathbf{A}_{-1} & \mathbf{0} & \ddots & & \\ \mathbf{0} & \mathbf{A}_{-1} & \ddots & & \\ \vdots & \ddots & \ddots & \ddots & \vdots \\ \mathbf{0} & \cdots & \mathbf{0} & \mathbf{A}_{-1} & \mathbf{0} \end{bmatrix}.$$

The MAP can be analyzed as described in Sect. 4 and the resulting quantities like moments, joint moments, lag k autocorrelation function or values of the joint densities can be used as input for the fitting algorithms introduced in the previous sections. In this way, a MAP with a smaller state space but a similar behavior is generated. In the original approach [70] the busy period of the queue is used to approximate the output process, however, this is just another measure to be approximated by a MAP.

In a QN of the type considered here, the output process might be split according to the routing probabilities and the input process of a queue may result from the superposition of several output or external arrival processes. Therefore, probabilistic splitting and superposition of MAPs have to be considered. We begin with splitting and assume that $\left(\mathbf{D}_0^{k-}, \mathbf{D}_1^{k-}\right)$ is a MAP approximating the output process of queue k and customers are routed with probability r_{kl} from k to l. Then $\left(\mathbf{D}_0^{k-} + (1-r_{kl})\mathbf{D}_1^{k-}, r_{kl}\mathbf{D}_1^{k-}\right)$ is the MAP approximation of the input process from queue k to l. Observe that splitting is an exact operation but since the MAP describing the output is usually only an approximation, the same holds for the MAP that describes the flow between both queues. Superposition of MAPs is introduced for two MAPs since the operation results in a new MAP, an arbitrary number of MAPs can be superposed. The superposition of two $\left(\mathbf{D}_0^1, \mathbf{D}_1^1\right)$ and $\left(\mathbf{D}_0^2, \mathbf{D}_1^2\right)$ results in a MAP $\left(\mathbf{D}_0^1 \oplus \mathbf{D}_0^2, \mathbf{D}_1^1 \oplus \mathbf{D}_1^2\right)$. The superposition is exact but the number of states of the resulting MAP equals the product of the state numbers of the superposed MAPs. However, the resulting MAP may be approximated by a MAP of lower order using the proposed fitting methods to generate the low order MAP.

With the basic approach to analyze a single $MAP/PH/1/N$ queue, an algorithm can be formulated that analyzes one queue after the other to compute the output process which is used as input process for other queues. This approach is iterated until the computed results remain stable. Usually decomposition based analysis approaches are efficient and have an adequate accuracy. However, the approximation error is often unknown and can only be computed if an exact analysis is possible or results can be compared with simulation results.

6.2 Modeling Reliability and Availability

PHDs and MAPs are an interesting model to describe failure and repair times or the duration of availability or unavailability intervals such that the resulting models can be mapped on Markov processes and analyzed numerically, which is important because often small probabilities have to be computed and can only hardly be estimated in a reliable way using simulation or approximation techniques. In the following we consider first the modeling of failure and repair times using PHDs and MAPs. Afterwards, a simple modeling approach is introduced where PHDs and MAPs are used to describe the behavior of components and the composed model is used to analyze the behavior of the whole system.

6.2.1 Modeling Failure and Repair Times

The analysis of failures and repairs is an important aspect of the analysis of technical systems. To build meaningful and realistic models it is important to have a precise representation of the time to failure and the required repair time of components and systems. Usually the model for the failure time is a part of a larger model that describes the consequences of the failure for the behavior of the whole system. Failures occur in any artificial system. We consider here the modeling of availability and unavailability intervals of components of large distributed systems. For such models a large amount of data is available. The failure trace archive [86] contains data describing the availability and unavailability phases of components for several large distributed systems.

Usually failure and also repair times are modeled by exponential, Weibull, log-normal or Gamma distributions [46]. Apart from the exponential distribution, which for real data often shows a bad fitting quality, the distributions can only be used in conjunction with simulation to perform system analysis. Furthermore, the distributions have only very few parameters such that the parameter fitting is simple and can be made in closed form from the available data. However, a small number of parameters also restricts the flexibility of a distribution and allows therefore only

Table 6.1 Comparison of the length of availability and unavailability intervals in the trace *g5k06* and the fitted distributions

	Availability intervals				Unavailability intervals			
	mean	*cv*	*skew*	*log-likel.*	*mean*	*cv*	*skew*	*log-likel.*
Trace	32.41	2.91	15.06	–	7.41	8.13	26.26	–
Exponential	32.41	1.00	2.00	−1318033	7.41	1.00	2.00	$-\infty$
Weibull	31.08	2.37	7.24	−1117466	2.36	3.97	16.74	−141961
Log-normal	84.62	18.67	6560.80	−1123175	1.51	11.20	1439.00	−59979
Gamma	32.08	1.72	3.43	−1134256	7.58	2.29	4.59	−247270
Hyper-Erlang-2	32.41	1.50	2.46	−1156754	7.41	3.02	4.59	−28913
Hyper-Erlang-5	32.41	2.71	7.86	−1106988	7.41	3.77	5.80	49743
Hyper-Erlang-10	32.41	2.70	7.66	−1105569	7.41	3.99	6.19	76251
Hyper-Erlang-15	32.41	2.60	6.63	−1104910	7.41	5.61	9.73	90099
APHD-3 (Momfit)	32.41	2.91	15.06	−1315858	7.41	8.13	26.25	−53919

a limited approximation of the measured data. Interestingly, PHDs have only rarely been used to model failure data. We present here the modeling of the duration of availability and unavailability phases of the components of a distributed system using data from the failure trace archive. More comprehensive results on the use of PHDs for failure and repair time modeling can be found in [34].

We model the data from the trace *Grid'5000 (g5k06)* which contains failure data of a large computer grid [85]. Table 6.1 shows the mean, coefficient of variation (*cv*) and the skewness (*skew*) of the trace and several distributions which have been computed to approximate the trace behavior. All values are given in *hours*. In the last column the value of the log-likelihood of the different distributions with respect to the trace is shown.

The exponential, Weibull, log-normal and Gamma distributions have already been presented in [86], the Hyper-Erlang distributions have been generated with the EM algorithm in Sect. 3.1.3 and are already presented in [34] and the distribution *APHD-3 (Momfit)* has been generated by fitting the parameters to match the first three moments using the approach presented in Sect. 3.2. It can be noticed that the exponential distribution is not appropriate to model the interval length of availability and unavailability intervals, whereas the Hyper-Erlang distributions provide in all cases a much better approximation, if the number of phases is large enough and an EM algorithm is used to compute the parameters. It can be seen that a moment fitting approach is not appropriate. Even if the first three moments are matched exactly by the distribution, the log likelihood value for the model of the availability intervals is smaller than the log likelihood value of all other distributions, except the exponential distribution. For the availability intervals, the Weibull, log-normal and Gamma distribution are similar in terms of the likelihood. However, the Hyper-Erlang distribution with 5 phases has a larger likelihood than the mentioned distributions but the value grows only slowly, if the number of phases of the Hyper-Erlang distribution is increased. For modeling the unavailability, Hyper-Erlang distributions are much better than any other distribution because the values of the log likelihood function are significantly larger and still grow with an increasing number of phases.

PHDs can be used if the length of availability and unavailability intervals are identically and independently distributed. This is often not the case in real systems. In many cases, consecutive intervals are correlated. This correlation can be included in the model by using a MAP rather than a PHD to describe the interval length. Some first results using MAPs to include correlation in failure time modeling are presented in [34].

6.2.2 Reliability and Availability Models

In the previous section it has been shown how failure times can be modeled with PHDs. Now we build models that contain components for which the failure times are described by PHDs. There is no established approach to specify reliability models with PHDs, although these models are widely used, some examples are given in [122, 126, 147].

As an example application we consider here a simple class of models with failures but without repairs. We will show that the overall model can be interpreted as PHD and the distribution of the time to failure can be computed as the inter-event time of this PHD. Assume that the system consists of K components with a failure time that is described by a PHD $(\boldsymbol{\pi}^k, \mathbf{D}_0^k)$ of order n^k for component k. Again the state space of the system can be structured into macro states. A global state is described by vector $\mathbf{b} \in \mathbb{B}^K$ where $\mathbf{b}(k) = 1$ indicates that component k is working. Function $up : \mathbb{B}^K \to \mathbb{B}$ classifies global states as up or down states. In an up state the system is working, in a down state the system is defect and does no longer work. Since the system is shut down in a down state, no more components can fail and it is sufficient to consider a single down state which is absorbing in the Markov process. We assume that the number of macro states $\mathbf{b}$ with $up(\mathbf{b}) = 1$ is N and denote macro states by $\mathbf{b}_I$ $(I = 1, \ldots, N)$. The number of states of the Markov process equals $n = 1 + \sum_{I=1}^{N} \prod_{k:\mathbf{b}_I(k)=1} n^k$. The absorbing down state is the last state and macro states are ordered in a decreasing order according to the number of working components in vector $\mathbf{b}$. Initially the system is in a state where all components are working, i.e. $\mathbf{b}_1 = \mathbb{1}^T$. Matrix $\mathbf{Q}$ can then be structured in submatrices $\mathbf{Q}_{\mathbf{b},\mathbf{b}'}$ $(\mathbf{b}, \mathbf{b}' \in \mathbb{B}^K$, $up(\mathbf{b}) = up(\mathbf{b}') = 1)$.

$$\mathbf{Q} = \begin{bmatrix} \mathbf{Q}_{\mathbf{b}_1,\mathbf{b}_1} & \cdots & \cdots & \mathbf{Q}_{\mathbf{b}_1,\mathbf{b}_N} & \mathbf{q}_{\mathbf{b}_1} \\ \mathbf{0} & \mathbf{Q}_{\mathbf{b}_2,\mathbf{b}_2} & \cdots & \mathbf{Q}_{\mathbf{b}_2,\mathbf{b}_N} & \mathbf{q}_{\mathbf{b}_2} \\ \vdots & \ddots & \ddots & \vdots & \vdots \\ \vdots & & \ddots & \mathbf{Q}_{\mathbf{b}_N,\mathbf{b}_N} & \mathbf{q}_{\mathbf{b}_N} \\ \mathbf{0} & \cdots & \cdots & \mathbf{0} & 0 \end{bmatrix}$$

If we assume that components fail independently, then the submatrices can be generated from the matrices of the PHDs as follows

$$\mathbf{Q}_{\mathbf{b}_I,\mathbf{b}_J} = \begin{cases} \bigoplus_{k:\mathbf{b}_I(k)=1} \mathbf{D}_0^k & \text{if } I = J \\ \mathbf{I}_{\prod_{l<k:\mathbf{b}_I(k)=1} n^l} \otimes \mathbf{d}_1^k \bigotimes \mathbf{I}_{\prod_{l>k:\mathbf{b}_I(k)=1} n^l} & \text{if } \mathbf{b}_I - \mathbf{e}_k = \mathbf{b}_J \text{ and} \\ \mathbf{0} & \text{otherwise} \end{cases}$$

$$\mathbf{q}_{\mathbf{b}_I} = \sum_{k:\mathbf{b}_I(k)=1, up(\mathbf{b}_I-\mathbf{e}_k)=0} \mathbb{1}_{\prod_{l<k:\mathbf{b}_I(k)=1} n^l} \otimes \mathbf{d}_1^k \otimes \mathbb{1}_{\prod_{l>k:\mathbf{b}_I(k)=1} n^l},$$

where $\mathbb{1}_n$ is the unit vector of order n. If we assume that the systems starts with new components, the initial vector equals $\boldsymbol{\pi} = [\otimes_{k=1}^K \boldsymbol{\pi}^k, \mathbf{0}, \ldots, \mathbf{0}, 0]$, i.e. the initial probabilities of the states in the first macro state are defined according to the initial distributions of the PHDs and the initial probability of all remaining states becomes 0. Now $(\boldsymbol{\pi}, \mathbf{Q})$ can be interpreted as a PHD and the inter-event time describes the time to failure of the system.

Example 6.3. If we consider a 2-out-3 system, then following macro states describe the system in an up state: $(1,1,1), (1,1,0), (1,0,1), (0,1,1)$. Matrix $\mathbf{Q}$ and initial vector $\boldsymbol{\pi}$ can then be computed as shown above if the PHDs describing the failure times of the three components are available.

6.3 Simulation Models

PHDs and MAPs do not belong to the distributions or stochastic processes that are commonly used in simulation [105] and are therefore not available as standard components in simulation tools [71, 92]. However, it is usually straightforward to integrate PHDs or MAPs into simulation software. Here we first introduce the basic approach to generate random numbers from PHDs or MAPs. Afterwards, we present some simulation models that include PHDs and MAPs.

6.3.1 *Generating Random Numbers from PHDs and MAPs*

A MAP is completely specified by the matrices $\mathbf{D}_0$ and $\mathbf{D}_1$. For simulation we define $\lambda(i) = -\mathbf{D}_0(i,i)$ the rate of the exponential distribution specifying the holding time in state i and two $n \times n$ matrices $\mathbf{P}_0$ and $\mathbf{P}_1$ with

$$\mathbf{P}_0(i,j) = \begin{cases} \lambda(i)^{-1}\mathbf{D}_0(i,j) & \text{for } i \neq j \\ 0 & \text{for } i = j \end{cases} \quad \text{and } \mathbf{P}_1(i,j) = \lambda(i)^{-1}\mathbf{D}_1(i,j).$$

Observe that by definition of a MAP $\lambda(i) > 0$ for all i. It follows that $(\mathbf{P}_0 + \mathbf{P}_1)\mathbb{1} = \mathbb{1}$. Define $\mathbf{p}_0 = \mathbf{P}_0\mathbb{1}$. For simulation the MAP is usually initialized according to the stationary distribution defined as the solution of $\boldsymbol{\psi}(\mathbf{D}_0 + \mathbf{D}_1) = \mathbf{0}$ and $\boldsymbol{\psi}\mathbb{1} = 1$. To simulate a MAP, random numbers have to be drawn from a uniform $[0,1)$

Algorithm 5 Generate_Random_Number_from_MAP$((\lambda(i))_{i=1,...,n}, \mathbf{P}_0, \mathbf{p}_0, \mathbf{P}_1, i)$

Input: MAP $(\mathbf{D}_0, \mathbf{D}_1)$, current state i;
Output: Time to the next event x and new state i;
 1: initialize $x = 0$ and $found = false$;
 2: **repeat**
 3:   $x = x + exp(\lambda(i))$;
 4:   $u = unif$;
 5:   **if** $u \leq \mathbf{p}_0(i)$ **then**
 6:     $sum = 0$;
 7:     $j = 1$;
 8:     **repeat**
 9:       $sum = sum + \mathbf{P}_0(i, j)$;
10:       $j = j + 1$;
11:     **until** $sum > u$;
12:   **else**
13:     $sum = \mathbf{p}_0(i)$;
14:     $j = 1$;
15:     **repeat**
16:       $sum = sum + \mathbf{P}_1(i, j)$;
17:       $j = j + 1$;
18:     **until** $sum > u$;
19:     $found = true$;
20:   **end if**
21:   $i = j$;
22: **until** $found == true$;
23: return x and i ;

---

distribution (denoted as *unif*) and from an exponential distribution with rate λ (denoted as $exp(\lambda)$). Procedures *unif* and $exp(\cdot)$ are commonly available in simulation software. Algorithm 5 generates a new random number from a MAP and returns, additionally, the new state of the MAP.

Before the procedure is called for the first time, the initial state i is initialized according to the stationary distribution defined in ψ. For each random number generation from the MAP, Algorithm 5 is called with the current state of the MAP and returns a new random number and a new state which is used for the next call of the function. The effort for random number generation depends on the number of iterations that are necessary to generate a new value and the number of iterations depends on the size of the elements in vector $\mathbf{p}_0$. The effort per iteration is proportional to the number of non-zero elements in a row of $\mathbf{D}_0$ and $\mathbf{D}_1$. Algorithm 5 can also be applied to generate random number from a PHD by interpreting the PHD $(\boldsymbol{\pi}, \mathbf{D}_0)$ as a MAP $(\mathbf{D}_0, \mathbf{d}_1\boldsymbol{\pi})$ (cf. Sect. 4.1.1).

The algorithm can be implemented in standard simulation tools that support the definition of modules, which is common in modern simulation software. An implementation in the network simulation tool OMNeT++ [71] is presented in [99].

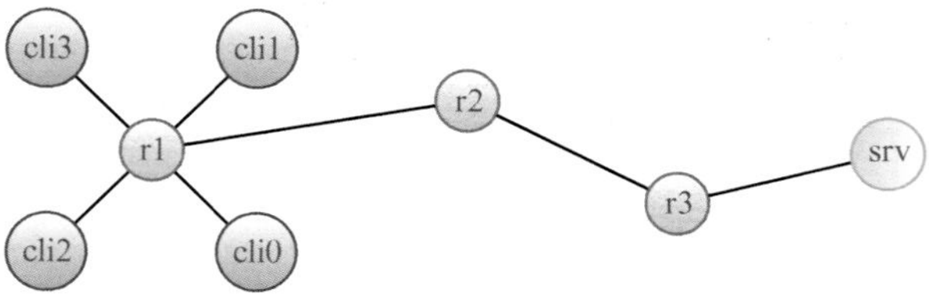

Fig. 6.3 Network topology of *NClients* example

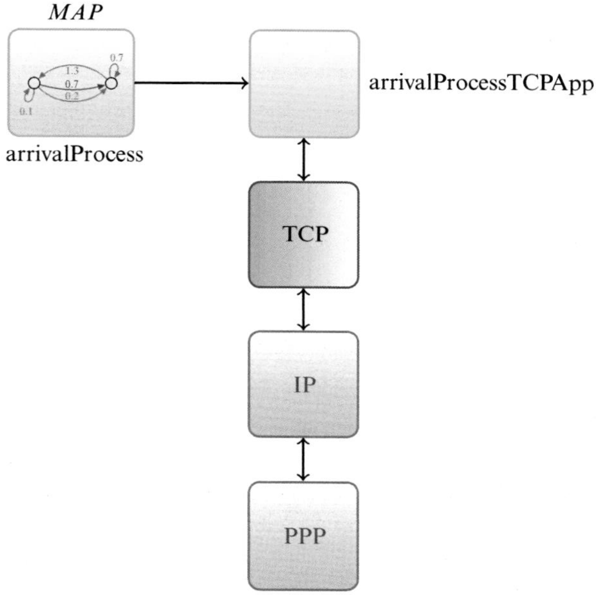

Fig. 6.4 Host from *NClients* example

6.3.2 Simulation Models with PHDs and MAPs

Since network traffic is known to exhibit dependencies and correlations, we present two models of computer networks as an application for PHDs and MAPs in simulation. We use the above mentioned network simulator OMNeT++ [71] in combination with the INET framework, that provides implementations of various network protocols. The models are both example models from the INET framework that we modified to account for random numbers from PHDs and MAPs using the *Arrival Process Module* from [99] (cf. Sect. 7.3).

The first example, called *NClients*, consists of four identical clients (*cli0* - *cli4*) that are connected to a server (*srv*) via several routers (*r1* - *r3*). The network layout is shown in Fig. 6.3. The inner view of the clients is sketched in Fig. 6.4. The host consists of modules that implement the transport and the network layer and two modules (*arrivalProcess* and *arrivalProcessTCPApp*) that are responsible for traffic

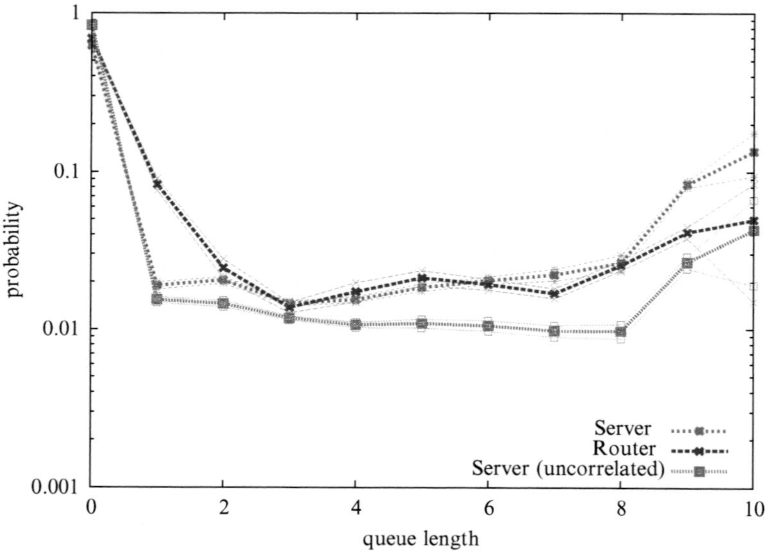

Fig. 6.5 Simulation results for the *NClients* example

generation. The module *arrivalProcess* simulates a MAP of order 4 and whenever the MAP generates an event, a message is sent to the module *arrivalProcessTCPApp* to trigger traffic generation, i.e. the module sends a request to the server. The size of the reply from the server is drawn from a hyper-exponential distribution consisting of a phase with a small rate and a phase with a large rate. In this way typical web browsing behavior is simulated where the users request various smaller files and some larger downloads from a web server.

During the simulation we measured the queue length distribution at the network interfaces of the different routers, the server and the clients. Figure 6.5 shows the distribution for the network interface of the server and for the router directly connected to the server resulting from 30 replications. As we can see both, router and server are busy due to the larger files they have to deliver. The thin lines show the 95% confidence intervals, that are very narrow except for the large queue length values. Note, that the confidence interval looks asymmetric in some cases due to the logarithmic scale of the y-axis. Figure 6.5 also shows a curve labeled with `Server(uncorrelated)`. This curve results from a simulation of the *NClients* model with uncorrelated arrivals, i.e. traffic generation is not triggered by a MAP but by the PHD defined by the stationary inter-event distribution of the MAP resulting in uncorrelated packet generation times. Comparison between the two curves for the server shows the impact that correlation can have on performance measures. With uncorrelated arrivals we have a larger probability for an empty queue while the probability for larger queue populations is smaller. This implies that one might underestimate the resource requirements of a system when neglecting correlation in a simulation model.

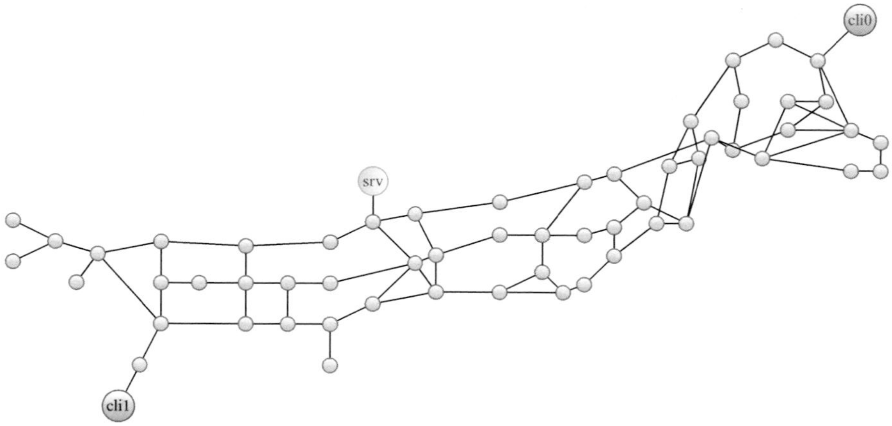

Fig. 6.6 Network topology of *FlatNet* example

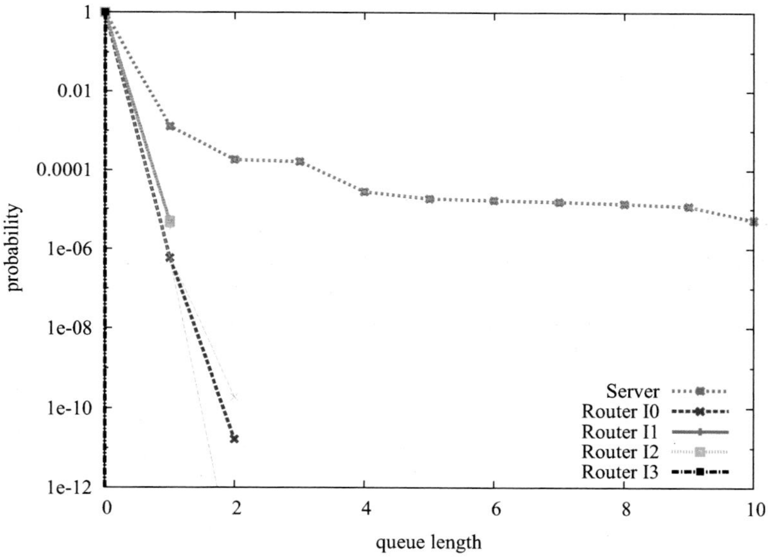

Fig. 6.7 Simulation results for the *FlatNet* example

In the previous example we simulated a rather small network. A larger network of routers is shown in Fig. 6.6. We connected two clients and a server to the network. The clients are configured as in the previous example, i.e. requests are generated by a MAP and the size of the requested file is modeled by a PHD. Figure 6.7 shows the queue length distribution and the 95 % confidence intervals of the network interfaces of the server and the router directly connected to it resulting from 30 replications. Since the packets can take different routes through the network the router interfaces are utilized differently, e.g. interface 3 is not used at all.

6.4 Concluding Remarks

In this section some examples for the use of PHDs and MAPs in stochastic models have been presented. Apart from classical queueing networks for which PHDs and MAPs originally have been developed, reliability models and network simulation models are introduced. Both examples show that PHDs and MAPs can be used in different application areas and in conjunction with different solution methods.

The presented examples are only a very small excerpt for the use of PHDs and MAPs in stochastic modeling. There are numerous other modeling formalisms like stochastic Petri Nets [64] or stochastic process algebras [52] that have been extended to include PHDs to model event times. For a detailed description of models that are mapped on Markov processes we refer to the literature [49, 151, 152]. However, apart from queueing networks, which naturally contain MAPs as arrival or service processes, we are not aware of other modeling formalisms where MAPs are fully integrated. This also holds for simulation models where PHDs are commonly applied [27, 81] but MAPs are rarely used.

PHDs and MAPs are often used in decomposition based approximate analysis [50, 70, 79]. The central idea of the approximate solution approaches is to approximate the output process of some complex subsystem, usually a subnet in a queueing network, by a PHD or MAP with less states. Often moments or joint moments of the output process of the original submodel are approximated by a PHD or MAP. Parameters of the PHD or MAP are computed with the methods presented for moment based fitting in the previous sections.

Performance and dependability analysis of computer and communication systems are the application areas that have been considered in this section. There are several other applications area where PHDs and MAPs have been applied. Typical examples are computational finance [8, 9, 17], where MAPs and PHDs are used in risk models, the analysis of manufacturing systems [43], where MAPs and PHDs are a common model for the load of production lines or inventory systems, the analysis of healthcare systems [55], where PHDs describe the time the patients stay in hospitals, or infection models [61], where PHDs model the duration of different phases of an infection.

Chapter 7
Software Tools

Various of the approaches and algorithms presented in the previous sections are available in software tools. The majority of the tools has been designed for parameter estimation of PHDs and MAPs and will be presented in Sects. 7.1 and 7.2, respectively. Of course, the resulting PHDs or MAPs will usually serve as input model (e.g. characterizing inter-arrival or service times) to some larger model which should be analyzed, either by simulation or by applying numerical techniques. The last section of this chapter deals with software to analyze these models.

7.1 Software Tools for Generating PH Distributions

Gfit [156] is a command line tool that implements an EM algorithm to fit hyper-Erlang distributions as described in Sect. 3.1.3. The tool can either fit a single HErD with a given number of branches and states per branch to a trace or find the best (i.e. the one with the largest likelihood) HErD for a given overall number of states. The latest version of the tool also supports uniform and logarithmic trace aggregation [139].

PhFit[1] [75] uses different distance measures like the relative entropy or the pdf (cdf) area distance to estimate the parameters of PHDs in series canonical form (cf. Sect. 3.1.5). To cope with heavy-tailed distributions *PhFit* can estimate the body and the tail separately and combine the two PHDs into one larger PHD. Additionally, *PhFit* is one of the few tools that can also estimate the parameters of discrete PHDs. *PhFit* provides a graphical user interface to set the parameters for the fitting procedure and provides plots of the empirical distribution and the resulting PHD.

[1] http://webspn.hit.bme.hu/~telek/tools.htm.

P. Buchholz et al., *Input Modeling with Phase-Type Distributions and Markov Models: Theory and Applications*, SpringerBriefs in Mathematics,
DOI 10.1007/978-3-319-06674-5_7,

HyperStar[2] [144] supports hyper-Erlang distributions as well as other mixtures of PHDs. *HyperStar* uses a cluster-based approach for fitting.

PHPACK[3] [134] is a command line tool that implements an EM algorithm to fit general APHDs and APHDs in series canonical form as described in Sect. 3.1.4. The tool can fit APHDs to either an unsorted trace consisting of a sequence of times or to a sorted weighted trace. *PHPACK* also provides procedures for computation of pdf, cdf, moments, and generation of weighted samples from some theoretical distributions, e.g. Weibull or Lognormal, which can be directly used for APHD fitting.

7.2 Software Tools for Generating MAPs

The *KPC-Toolbox*[4] [41] is a collection of functions for MATLAB that implement the compositional approach from Sect. 5.1.2, i.e. the algorithm fits several smaller MAPs of order 2 according to empirical moments and combines them into one large MAP using Kronecker operations. The *KPC-Toolbox* also contains an approach for automatic selection of the number of states of the resulting MAP using the Bayesian Information Criterion (BIC) [150].

ProFiDo (Processes Fitting Toolkit Dortmund)[5] [12, 13] is a general framework for the parameter estimation of stochastic processes with the main focus on PHDs and MAPs. As described in Sect. 5, it is often necessary and helpful to combine different fitting approaches for PHDs and MAPs. For example the two-step MAP fitting approaches expand a PHD that can be fitted with any of the available tools for PHDs. The runtime of EM algorithms for MAPs can be reduced significantly, if they are initialized with a good MAP as starting point, e.g. with one that has been generated with any of the other more efficient but probably less accurate approaches. Unfortunately, most of the existing fitting tools are not compatible due to different input and output formats. *ProFiDo* has been developed to diminish these obstacles and allows for an easy specification of workflows comprising different fitting tools. Workflows can be specified in a graphical manner and consist of so-called job nodes representing different tools for parameter estimation or result visualization. Job nodes are connected by directed arcs depicting the flow of information, i.e. the resulting description of a fitted distribution or process from the first node is used as input to the second node. *ProFiDo* includes various tools for parameter estimation which have been partially described in the previous paragraphs already. For the parameter estimation of PHDs two approaches are available, namely *Gfit* and an implementation of the least squares based technique presented in Sect. 3.2.2

[2] http://www.mi.fu-berlin.de/inf/groups/ag-tech/projects/HyperStar/index.html.

[3] http://www.rel.hiroshima-u.ac.jp/~okamu/PHPACK/.

[4] http://www.cs.wm.edu/MAPQN/kpctoolbox.html.

[5] http://ls4-www.cs.tu-dortmund.de/profido.

[32]. For MAP fitting *ProFiDo* offers two-step fitting according to joint moments (cf. Sect. 5.3.1) and autocorrelation coefficients (cf. Sect. 5.3.2) and EM algorithms as described in Sect. 5.2. In addition, tools to visualize density, distribution, autocorrelation and (joint) moments are available, traces can be generated from the fitted process descriptions, the queueing behavior can be compared and statistical tests are supported. Moreover, fitting algorithms for other stochastic processes based on Autoregressive Moving Average models are also offered by *ProFiDo*.

To ensure interoperability between these different tools *ProFiDo* uses an XML format for process description and scripts to convert from a tool's native format to the XML description and vice versa.

ProFiDo is linked to different tools for the analysis of models with PHDs and MAPs which will be described in the next section.

BUTools[6] is a collection of scripts for Mathematica and Matlab/Octave that cover functions for PHDs and MAPs and their generalizations matrix exponential distributions and Rational Arrival Processes. Aside from scripts to compute (joint) moments, autocorrelations or to compute canonical representations, exact moment matching for PHDs and MAPs is supported using the second approach from Sect. 3.2.1 for PHDs and a generalization of these ideas for MAPs [155].

7.3 Software Tools for Analyzing Models with PHDs and MAPs

In Chap. 6 several different types of models for which PHDs and MAPs are commonly used have been reviewed. Depending on the type of model different tools for the analysis are available.

For the network simulation tool OMNeT++ two libraries/modules are available that support random number generation from PHDs or MAPs. The *Arrival Process Module*[7] [99] can generate random numbers from several stochastic processes including MAPs. PHDs are supported when they are represented as a MAP. The module basically implements Algorithm 5 from Sect. 6.3. The *Arrival Process Module* is linked to the aforementioned toolset *ProFiDo* and can directly import *ProFiDo*'s XML format for process description such that including PHDs or MAPs fitted with *ProFiDo* into an OMNeT++ model is possible without additional effort.

If only uncorrelated random numbers from a PHD are required *libphrng* [143] is available which can be used in OMNeT++ models with only little effort as well. *libphrng* is part of the *BUTools* package and implements efficient algorithms for random number generation from [81].

[6] http://webspn.hit.bme.hu/~telek/tools/butools/butools.html.

[7] http://ls4-www.cs.tu-dortmund.de/profido.

If the MAP or PHD is used as part of a larger Markovian model, solvers for Markov Chains like *NSOLVE*[8] can be used to analyze the model. *NSOLVE* is based on a compositional description of Markov models where PHDs and MAPs may be used as arrival or service processes.

Queueing system of the type *MAP*/*MAP*/1 can be analyzed with the tool *Q-MAM*[9] which is a MATLAB toolbox implementing several up to date matrix analytical solvers.

[8] http://ls4-www.cs.tu-dortmund.de/download/buchholz/struct-matrix-market.html.

[9] http://win.ua.ac.be/~vanhoudt.

Chapter 8
Conclusion

This book collects available results on PHDs and MAPs and, in particular, it presents several of the available methods to determine the parameters of a PHD or MAP in order to capture the behavior of a real system described in form of some measurements. Our goal was to provide an application oriented presentation that helps to apply available techniques in practical modeling. We hope that the book helps a reader to solve her or his modeling problems when complex processes, which are common in many systems, have to be modeled and analyzed.

PHDs and MAPs have a great potential in describing real processes but the parameter fitting problem is still a complex optimization problem and we cannot claim that the resulting problems are all solved by the available methods. Nevertheless, practical experience shows that even multimodal distributions can be closely approximated by PHDs for which the parameters have been determined using one of the methods presented in this book. The situation for modeling stochastic processes is less advanced, since the parameter fitting problem for MAPs is still a challenge. However, currently available methods often give good results and require an acceptable effort.

The field of Markov models to approximate distributions and stochastic processes is very wide and we cannot provide an exhaustive overview of the whole area. There are some other overview papers [2, 3, 62, 76] and a huge number of research papers spread over different fields like computer science, applied mathematics, statistics and operations research. We hope that we mentioned the most important application oriented papers in the text, but for sure we forgot some.

The large number of publications on PHDs and MAPs shows a growing interest in these models types. In the future new and more efficient methods for parameter fitting will be developed and become available in software tools. There are several developments which look very interesting but have not been included in this book, partially because the length of the text is limited and partially the approaches are not developed far enough. As examples we would like to mention the combination of PHDs and autoregressive processes [98], the parameter fitting for MMAPs and BMAPs [39, 65, 97] and the use of ME distributions [67, 111] and (M)RAPs [5, 38] rather than PHDs and (M)MAPs.

P. Buchholz et al., *Input Modeling with Phase-Type Distributions and Markov Models: Theory and Applications*, SpringerBriefs in Mathematics,
DOI 10.1007/978-3-319-06674-5_8, © Peter Buchholz, Jan Kriege, Iryna Felko 2014

References

1. Alfa, A.S., Neuts, M.F.: Modelling vehicular traffic using the discrete time Markovian arrival process. Transport. Sci. **29**(2), 109–117 (1995)
2. Artalejo, J.R., Gomez-Corral, A., He, Q.M.: Markovian arrivals in stochastic modelling: a survey and some new results. SORT **34**(2), 101–156 (2010)
3. Asmussen, S.: Phase-type distributions and related point processes: fitting and recent advances. In: Chakravarthy, S.R., Alfa, A.S. (eds.) Matrix-Analytic Methods in Stochastic Models. Lecture Notes in Pure and Applied Mathematics, pp. 137–149. Dekker, New York (1997)
4. Asmussen, S.: Applied Probability and Queues. Springer, New York (2003)
5. Asmussen, S., Bladt, M.: Point processes with finite-dimensional conditional probabilities. Stoch. Process. Their Appl. **82**, 127–142 (1999)
6. Asmussen, S., Nerman, O., Olsson, M.: Fitting phase-type distributions via the EM-algorithm. Scand. J. Stat. **23**(4), 419–441 (1996)
7. Atkinson, K.A.: An Introduction to Numerical Analysis, 2nd edn. Wiley, New York (1989)
8. Badescu, A.L., Drekic, S., Landriault, D.: Analysis of a threshold divided strategy for a MAP risk model. Scand. Actuar. J. **4**, 227–247 (2007)
9. Badescu, A.L., Cheung, E.K., Landriault, D.: Dependent risk models with bivariate phase-type distributions. J. Appl. Probab. **46**(1), 113–131 (2009)
10. Balsamo, S., de Nitto Persone, V., Onvural, R.: Analysis of Queueing Networks with Blocking. International Series on Operations Research and Management Science. Kluwer Academic Publishers, Boston (2001)
11. Bause, F., Buchholz, P., Kriege, J.: A comparison of Markovian arrival processes and ARMA/ARTA processes for the modelling of correlated input processes. In: Proceedings of the Winter Simulation Conference (2009)
12. Bause, F., Buchholz, P., Kriege, J.: ProFiDo: the processes fitting toolkit Dortmund. In: Proceedings of the 7th International Conference on Quantitative Evaluation of Systems (QEST 2010), pp. 87–96. IEEE Computer Society, Williamsburg (2010)
13. Bause, F., Gerloff, P., Kriege, J.: ProFiDo: a toolkit for fitting input models. In: Müller-Clostermann, B., Echtle, K., Rathgeb, E.P. (eds.) Proceedings of the 15th International GI/ITG Conference on Measurement, Modelling, and Evaluation of Computing Systems and Dependability and Fault Tolerance. Lecture Notes in Computer Science, vol. 5987, pp. 311–314. Springer, Berlin (2010)
14. Biller, B., Gunes, Ç.: Introduction to simulation input modeling. In: Johansson, B., Jain, S., Montoya-Torres, J., Hugan, J., Yücesan, E. (eds.) Proceedings of the Winter Simulation Conference (WSC), pp. 49–58 (2010)

P. Buchholz et al., *Input Modeling with Phase-Type Distributions and Markov Models: Theory and Applications*, SpringerBriefs in Mathematics,
DOI 10.1007/978-3-319-06674-5,

15. Bilmes, J.: A gentle tutorial on the EM algorithm and its application to parameter estimation for Gaussian mixture and hidden Markov models. Technical Report TR-97-021, University of Berkeley (1997)
16. Bini, D.A., Latouche, G., Meini, B.: Numerical Methods for Structured Markov Chains. Oxford Science Publications, Oxford (2005)
17. Bladt, M.: A review on phase-type distributions and their use in risk theory. Astin Bull. **35**(1), 145–161 (2005)
18. Bobbio, A., Cumani, A.: ML estimation of the parameters of a PH distribution in triangular canonical form. In: Balbo, G., Serazzi, G. (eds.) Computer Performance Evaluation, pp. 33–46. Elsevier, Amsterdam (1992)
19. Bobbio, A., Telek, M.: Parameter estimation of phase type distributions. Technical Report R.T.423, Instituto Elettrotechnico Nazional Galileo Ferraris (1997)
20. Bobbio, A., Horváth, A., Scarpa, M., Telek, M.: Acyclic discrete phase type distributions: properties and a parameter estimation algorithm. Perform. Eval. **54**(1), 1–32 (2003)
21. Bobbio, A., Horváth, A., Telek, M.: Matching three moments with minimal acyclic phase type distributions. Stoch. Model. **21**(2–3), 303–326 (2005)
22. Bodrog, L., Heindl, A., Horváth, G., Telek, M., Horváth, A.: A Markovian canonical form of second-order matrix-exponential processes. Eur. J. Oper. Res. **160**(1), 51–68 (2008)
23. Bodrog, L., Heindl, A., Horváth, G., Telek, M., Horváth, A.: Current results and open questions on PH and MAP characterization. In: Bini, D., Meini, B., Ramaswami, V., Remiche, M., Taylor, P. (eds.) Numerical Methods for Structured Markov Chains, No. 07461 in Dagstuhl Seminar Proceedings (2008)
24. Bodrog, L., Buchholz, P., Kriege, J., Telek, M.: Canonical form based MAP(2) fitting. In: Proceedings of the 7th International Conference on the Quantitative Evaluation of Systems (QEST), pp. 107–116. IEEE Computer Society, Williamsburg (2010)
25. Breuer, L.: An EM algorithm for batch Markovian arrival processes and its comparison to a simpler estimation procedure. Ann. OR **112**(1–4), 123–138 (2002)
26. Breuer, L., Kume, A.: An EM algorithm for Markovian arrival processes observed at discrete times. In: Müller-Clostermann, B., Echtle, K., Rathgeb, E. (eds.) Measurement, Modelling, and Evaluation of Computing Systems and Dependability and Fault Tolerance. Lecture Notes in Computer Science, vol. 5987, pp. 242–258. Springer, Berlin (2010)
27. Brickner, C., Indrawan, D., Williams, D., Chakravarthy, S.R.: Simulation of a stochastic model for a service system. In: Johansson, B., Jain, S., Montoya-Torres, J., Hugan, J., Yücesan, E. (eds.) Proceedings of the Winter Simulation Conference (WSC), pp. 1636–1647 (2010)
28. Buchholz, P.: A class of hierarchical queueing networks and their analysis. Queueing Syst. **15**(1), 59–80 (1994)
29. Buchholz, P.: Exact and ordinary lumpability in finite Markov chains. J. Appl. Probab. **31**, 59–75 (1994)
30. Buchholz, P.: Structured analysis approaches for large Markov chains. Appl. Numer. Math. **31**(4), 375–404 (1999)
31. Buchholz, P.: An EM-algorithm for MAP fitting from real traffic data. In: Kemper, P., Sanders, W.H. (eds.) Computer Performance Evaluation/TOOLS. Lecture Notes in Computer Science, vol. 2794, pp. 218–236. Springer, New York (2003)
32. Buchholz, P., Kriege, J.: A heuristic approach for fitting MAPs to moments and joint moments. In: Proceedings of the 6th International Conference on the Quantitative Evaluation of Systems (QEST), pp. 53–62. IEEE Computer Society, Budapest (2009)
33. Buchholz, P., Kriege, J.: Equivalence transformations for acyclic phase type distributions. Technical Report 827, Department of Computer Science, TU Dortmund. http://www.cs.uni-dortmund.de/nps/de/Forschung/Publikationen/Graue_Reihe1/Ver__ffentlichungen_2009/827.pdf (2009)
34. Buchholz, P., Kriege, J.: Markov modeling of availability and unavailability data. In: Proceedings of the 10th European Dependable Computing Conference (EDCC), IEEE Computer Society, Newcastle upon Tyne (2014)

35. Buchholz, P., Panchenko, A.: An EM algorithm for fitting of real traffic traces to PH-distribution. In: Proceedings of the International Conference on Parallel Computing in Electrical Engineering, PARELEC, pp. 283–288. IEEE Computer Society, Dresden (2004)
36. Buchholz, P., Telek, M.: Stochastic Petri nets with matrix exponentially distributed firing times. Perform. Eval. **67**(12), 1373–1385 (2010)
37. Buchholz, P., Telek, M.: Rational arrival processes associated to labelled Markov processes. J. Appl. Probab. **49**(1), 40–59 (2012)
38. Buchholz, P., Telek, M.: On minimal representations of rational arrival processes. Ann. Oper. Res. **202**(1), 35–58 (2013)
39. Buchholz, P., Kemper, P., Kriege, J.: Multi-class Markovian arrival processes and their parameter fitting. Perform. Eval. **67**(11), 1092–1106 (2010)
40. Buchholz, P., Felko, I., Kriege, J.: Transformation of acyclic phase type distributions for correlation fitting. In: Proceedings of the Analytical and Stochastic Modeling Techniques and Applications (ASMTA). Lecture Notes in Computer Science, pp. 96–111. Springer, Berlin (2013)
41. Casale, G., Zhang, E.Z., Smirni, E.: KPC-toolbox: simple yet effective trace fitting using Markovian arrival processes. In: Proceedings of the 5th International Conference on the Quantitative Evaluation of Systems (QEST), pp. 83–92. IEEE Computer Society, St. Malo (2008)
42. Casale, G., Zhang, E.Z., Smirni, E.: Trace data characterization and fitting for Markov modeling. Perform. Eval. **67**(2), 61–79 (2010)
43. Ching, W.K.: Iterative Methods for Queuing and Manufacturing Systems. Monographs in Mathematics. Springer, London (2001)
44. Collection of availability traces. http://www.cs.illinois.edu/~pbg/availability/
45. Cox, D.R.: A use of complex probabilities in the theory of stochastic processes. Math. Proc. Camb. Phil. Soc. **51**, 313–319 (1955)
46. Crowder, M.J., Kimber, A.C., Smith, R.L., Sweeting, T.J.: Statistical Analysis of Reliability Data. CRC Press, Boca Raton (1994)
47. Cumani, A.: On the canonical representation of homogeneous Markov processes modeling failure-time distributions. Micorelectron. Reliab. **22**(3), 583–602 (1982)
48. Dayar, T.: On moments of discrete phase-type distributions. In: Bravetti, M., Kloul, L., Zavattaro, G. (eds.) Proceedings of the EPEW/WS-FM. Lecture Notes in Computer Science, vol. 3670, pp. 51–63. Springer, New York (2005)
49. Dayar, T.: Analyzing Markov Chains Using Kronecker Products. Briefs in Mathematics. Springer, New York (2012)
50. Dayar, T., Meriç, A.: Kronecker representation and decompositional analysis of closed queueing networks with phase-type service distributions and arbitrary buffer sizes. Ann. OR **164**(1), 193–210 (2008)
51. Dempster, A., Laird, N., Rubin, D.: Maximum likelihood from incomplete data via the EM algorithm. J. R. Stat. Soc. Ser. B **39**(1), 1–38 (1977)
52. El-Rayes, A., Kwiatkowska, M., Norman, G.: Solving infinite stochastic process algebra models through matrix-geometric methods. In: Hillston, J., Silva, M. (eds.) Proceedings of the 7th Process Algebras and Performance Modelling Workshop, pp. 41–62 (1999)
53. Erlang, A.K.: Solution of some problems in the theory of probabilities of significance in automatic telephone exchanges. Elektrotkeknikeren **13**, 5–13 (1917)
54. Fackrell, M.: Characterization of matrix-exponential distributions. Ph.D. thesis, School of Applied Mathematics, The University of Adelaide (2003)
55. Fackrell, M.: Modelling healthcare systems with phase-type distributions. Health Care Manag. Sci. **12**, 11–26 (2009)
56. Failure trace archive. http://fta.scem.uws.edu.au/
57. Fang, Y.: Hyper-Erlang distribution model and its application in wireless mobile networks. Wirel. Netw. **7**(3), 211–219 (2001)
58. Feldmann, A., Whitt, W.: Fitting mixtures of exponentials to long-tail distributions to analyze network performance models. Perform. Eval. **31**, 245–279 (1998)

59. Fischer, W., Meier-Hellstern, K.S.: The Markov-modulated Poisson process (MMPP) cookbook. Perform. Eval. **18**(2), 149–171 (1993)
60. Fox, B.L., Glynn, P.W.: Computing Poisson probabilities. Commun. ACM. **31**(4), 440–445 (1988)
61. Garg, L., Masala, G., McClean, S.I., Micocci, M., Cannas, G.: Using phase type distributions for modelling HIV disease progression. In: Proceedings of the 25th International Symposium on Computer-Based Medical Systems (CBMS), pp. 1–4. IEEE, Computer Society (2012)
62. Gerhardt, I., Nelson, B.L.: On capturing dependence in point processes: matching moments and other techniques. Technical Report, Northwestern University (2009)
63. Goseva-Popstojanova, K., Trivedi, K.S.: Effects of failure correlation on software in operation. In: Proceedings of the 2000 Pacific Rim International Symposium on Dependable Computing (PRDC), pp. 69–76. IEEE Computer Society, Los Angeles (2000)
64. Haddad, S., Moreaux, P., Chiola, G.: Efficient handling of phase-type distributions in generalized stochastic Petri nets. In: Azéma, P., Balbo, G. (eds.) Proceedings of the 18th International Conference on ICATPN. Lecture Notes in Computer Science, vol. 1248, pp. 175–194. Springer, Berlin (1997)
65. He, Q.M., Neuts, M.: Markov arrival processes with marked transitions. Stoch. Process. Their Appl. **74**, 37–52 (1998)
66. He, Q.M., Zhang, H.: A note on unicyclic representations of phase type distributions. Stoch. Model. **21**, 465–483 (2005)
67. He, Q.M., Zhang, H.: On matrix exponential distributions. Adv. Appl. Probab. **39**(1), 271–292 (2007)
68. Heckmüller, S., Wolfinger, B.E.: Using load transformations for the specification of arrival processes in simulation and analysis. Simulation **85**(8), 485–496 (2009)
69. Heindl, A.: Inverse characterization of hyperexponential MAP(2)s. In: Proceedings of the Analytical and Stochastic Modelling Techniques and Applications (ASMTA), pp. 183–189 (2004)
70. Heindl, A., Telek, M.: Output models of MAP/PH/1(/K) queues for an efficient network decomposition. Perform. Eval. **49**(1/4), 321–339 (2002)
71. Hornig, R., Varga, A.: An Overview of the OMNeT++ Simulation Environment. In: Proceedings of 1st International Conference on Simulation Tools and Techniques for Communications, Networks and Systems (SIMUTools) (2008)
72. Heindl, A., Mitchell, K., van de Liefvoort, A.: Correlation bounds for second-order MAPs with application to queueing network decomposition. Perform. Eval. **63**(6), 553–577 (2006)
73. Heindl, A., Horváth, G., Gross, K.: Explicit inverse characterizations of acyclic MAPs of second order. In: Horváth, A., Telek, M. (eds.) Proceedings of the 3rd European Performance Engineering Workshop: EPEW. Lecture Notes in Computer Science, vol. 4054, pp. 108–122. Springer, Berlin (2006)
74. Horváth, A., Telek, M.: Approximating heavy tailed behavior with phase type distributions. In: Proceedings of the 3rd International Conference on Matrix-Analytic Methods in Stochastic Models. Leuven, Belgium (2000)
75. Horváth, A., Telek, M.: PhFit: a general purpose phase type fitting tool. In: Proceedings of the Performance Tools 2002. Lecture Notes in Computer Science, vol. 2324, pp. 82–91. Springer, Berlin (2002)
76. Horváth, A., Telek, M.: Markovian modeling of real data traffic: Heuristic phase type and MAP fitting of heavy tailed and fractal like samples. In: Calzarossa, M.C., Tucci, S. (eds.) Proceedings of the Performance 2002. Lecture Notes in Computer Science, vol. 2459, pp. 405–434. Springer, Berlin (2002)
77. Horváth, A., Telek, M.: Matching more than three moments with acyclic phase type distributions. Stoch. Model. **23**, 167–194 (2007)
78. Horváth, G., Telek, M.: On the canonical representation of phase type distributions. Perform. Eval. **66**, 396–409 (2009)
79. Horváth, A., Horváth, G., Telek, M.: A traffic based decomposition of two-class queueing network with priority service. Comput. Netw. **53**(8), 1235–1248 (2009)

80. Horváth, A., Rácz, S., Telek, M.: Moments characterization of order 3 matrix exponential distributions. In: Al-Begain, K., Fiems, D., Horváth, G. (eds.) Proceedings of the Analytical and Stochastic Modeling Techniques and Applications (ASMTA). Lecture Notes in Computer Science, vol. 5513, pp. 174–188. Springer, Berlin (2009)
81. Horváth, G., Reinecke, P., Telek, M., Wolter, K.: Efficient generation of PH-distributed random variates. In: Al-Begain, K., Fiems, D., Vincent, J.M. (eds.) Proceedings of the Analytical and Stochastic Modeling Techniques and Applications (ASMTA). Lecture Notes in Computer Science, vol. 7314, pp. 271–285. Springer, Berlin (2012)
82. Horváth, G., Telek, M., Buchholz, P.: A MAP fitting approach with independent approximation of the inter-arrival time distribution and the lag-correlation. In: Proceedings of the 2nd International Conference on the Quantitative Evaluation of Systems (QEST), pp. 124–133. IEEE CS Press, Torino (2005)
83. Ide, I.: Superposition of interrupted Poisson processes and its application to packetized voice multiplexers. In: Proceedings of the International Teletraffic Congress (ITC12) (1988)
84. The internet traffic archive. http://ita.ee.lbl.gov/
85. Iosup, A., Jan, M., Sonmez, O., Epema, D.H.: On the dynamic resource availability in grids. In: Proceedings of the 8th IEEE/ACM International Conference on Grid Computing (2007)
86. Javadi, B., Kondo, D., Iosup, A., Epema, D.H.J.: The failure trace archive: enabling the comparison of failure measurements and models of distributed systems. J. Parallel Distrib. Comput. **73**(8), 1208–1223 (2013)
87. Johnson, M.: Selecting parameters of phase distributions: combining nonlinear programming, heuristics, and Erlang distributions. INFORMS J. Comput. **5**(1), 69–83 (1993)
88. Johnson, M., Taaffe, M.: Matching moments to phase distributions: mixtures of Erlang distributions of common order. Stoch. Model. **4**(5), 711–743 (1989)
89. Johnson, M., Taaffe, M.: Matching moments to phase distributions: nonlinear programming approaches. Stoch. Model. **2**(6), 259–281 (1990)
90. Jordan, M.I., Jacobs, R.A.: Hierarchical mixtures of experts and the EM algorithm. Neural Comput. **6**(2), 181–214 (1994)
91. Kawanishi, K.: On the counting process for a class of Markovian arrival processes with an application to a queueing system. Queueing Syst. **49**, 93–122 (2005)
92. Kelton, W.D., Sadowski, R.P., Sadowski, D.A.: Simulation with Arena, 4th edn. McGraw-Hill, New York (2007)
93. Kemeny, J.G., Snell, J.L.: Finite Markov Chains, repr edn. University Series in Undergraduate Mathematics. VanNostrand, New York (1969)
94. Khayari, R.E.A., Sadre, R., Haverkort, B.: Fitting world-wide web request traces with the EM-algorithm. Perform. Eval. **52**, 175–191 (2003)
95. Kleinrock, L.: Queueing Systems, vol. 1. Wiley, New York (1975)
96. Kleinrock, L.: Queueing Systems, vol. 2. Wiley, New York (1976)
97. Klemm, A., Lindemann, C., Lohmann, M.: Modeling IP traffic using the batch Markovian arrival process. Perform. Eval. **54**(2), 149–173 (2003)
98. Kriege, J., Buchholz, P.: Correlated phase-type distributed random numbers as input models for simulations. Perform. Eval. **68**(11), 1247–1260 (2011)
99. Kriege, J., Buchholz, P.: Simulating stochastic processes with OMNeT++. In: Liu, J., Quaglia, F., Eidenbenz, S., Gilmore, S. (eds.) Proceedings of the 4th International ICST Conference on Simulation Tools and Techniques (SimuTools'11), pp. 367–374. ICST/ACM, Brussels (2011)
100. Krijnen, W.P.: Convergence of the sequence of parameters generated by alternating least squares algorithms. Comput. Stat. Data Anal. **51**, 481–489 (2006)
101. Kuczura, A.: The interrupted Poisson process as an overflow process. The Bell Syst. Tech. J. **52**(3), 437–448 (1973)
102. Latouche, G.: A phase-type semi-Markov point process. SIAM J. Algebr. Discrete Meth. **3**, 77–90 (1982)
103. Latouche, G., Ramaswami, V.: Introduction to Matrix Analytic Methods in Stochastic Modeling. ASA-SIAM Series on Statistics and Applied Probability. Society for Industrial and Applied Mathematics, Philadelphia (1987)

104. Latouche, G., Ramaswami, V.: Introduction to Matrix Analytic Methods in Stochastic Modeling. Society for Industrial and Applied Mathematics, Philadelphia (1999)
105. Law, A.M., Kelton, W.D.: Simulation Modeling and Analysis, 3rd edn. McGraw-Hill, Boston (2000). ISBN 0-07-059292-6
106. Law, A.M., McComas, M.G.: ExpertFit distribution-fitting software: how the ExpertFit distribution-fitting software can make your simulation models more valid. In: Chick, S.E., Sanchez, P.J., Ferrin, D.M., Morrice, D.J. (eds.) Proceedings of the Winter Simulation Conference, pp. 169–174. ACM, Berlin (2003)
107. Lawson, C.L., Hanson, B.J.: Solving Least Squares Problems. Prentice-Hall, Englewood Cliffs (1974)
108. Lazowska, E.D., Zahorjan, J., Graham, G.S., Sevcik, K.C.: Quantitative system performance-computer system analysis using queueing network models. Prentice Hall, Upper Saddle River (1984)
109. Leland, W.E., Taqqu, M.S., Willinger, W., Wilson, D.V.: On the self-similar nature of ethernet traffic (extended version). IEEE/ACM Trans. Netw. **2**(1), 1–15 (1994)
110. van de Liefvoort, A.: The moment problem for continuous distributions. Technical Report WP-CM-1990-02, University of Missouri, Kansas City (1990)
111. Lipsky, L.: Queueing Theory: A Linear Algebraic Approach. Springer, New York (2008)
112. Loan, C.F.: The ubiquitous Kronecker product. J. Comput. Appl. Math. **123**(1–2), 85–100 (2000)
113. Lucantoni, D.M.: New results on the single server queue with a batch Markovian arrival process. Stoch. Model. **7**(1), 1–46 (1991)
114. Lucantoni, D.M.: The BMAP/G/1 queue: a tutorial. In: Donatiello, L., Nelson, R.D. (eds.) Performance/SIGMETRICS Tutorials. Lecture Notes in Computer Science, vol. 729, pp. 330–358. Springer, Berlin (1993)
115. Lucantoni, D.M., Meier-Hellstern, K.S., Neuts, M.F.: A single-server queue with server vacations and a class of non-renewal arrival processes. Adv. Appl. Probab. **22**(3), 676–705 (1990)
116. Maier, R.S., O'Cinneide, C.A.: A closure characterisation of phase-type distributions. J. Appl. Probab. **29**(1), 92–103 (1992)
117. McLachlan, G.J., Krishnan, T.: The EM Algorithm and Extensions. Wiley, Hoboken (1997)
118. Mészáros, A., Telek, M.: A two-phase MAP fitting method with APH interarrival time distribution. In: Proceedings of the Winter Simulation Conference. ACM, Berlin (2012)
119. Meyer, C.D.: Matrix Analysis and Applied Linear Algebra. Society for Industrial and Applied Mathematics, Philadelphia (2004)
120. Minin, V.N., Suchard, M.A.: Counting labeled transitions in continuous-time Markov models of evolution. J. Math. Biol. **56**, 391–412 (2008)
121. Mocanu, S., Commault, C.: Sparse representations of phase-type distributions. Stoch. Model. **15**, 759–778 (1999)
122. Montoro-Cazorla, D., Pérez-Ocón, R.: A maintenance model with failures and inspection following Markovian arrival processes and two repair modes. Eur. J. Oper. Res. **186**(2), 694–707 (2008)
123. Narayana, S., Neuts, M.: The first two moments matrices of the counts for the Markovian arrival process. Stoch. Model. **8**, 694–707 (1992)
124. Neuts, M.F.: A versatile Markovian point process. J. Appl. Probab. **16**, 764–779 (1979)
125. Neuts, M.F.: Matrix-geometric solutions in stochastic models. Johns Hopkins University Press, Baltimore (1981)
126. Neuts, M.F., Meier, K.S.: On the use of phase type distributions in reliability modelling of systems with two components. OR Spectr. **2**(4), 227–234 (1981)
127. Nielsen, B.F.: Lecture notes on phase-type distributions for 02407 stochastic processes. http://www2.imm.dtu.dk/courses/02407/ (2012)
128. Nightingale, E.B., Douceur, J.R., Orgovan, V.: Cycles, cells and platters: an empirical analysis of hardware failures on a million consumer PCs. In: Kirsch, C.M., Heiser, G. (eds.) Proceedings of the EuroSys, pp. 343–356. ACM, Salzburg (2011)

129. O'Cinneide, C.A.: On non-uniqueness of representations of phase-type distributions. Stoch. Model. **5**, 247–259 (1989)
130. O'Cinneide, C.A.: Characterization of phase-type distributions. Stoch. Model. **6**, 1–57 (1990)
131. O'Cinneide, C.A.: Phase type distributions and invariant polytopes. Adv. Appl. Prob. **23**, 515–535 (1991)
132. O'Cinneide, C.A.: Phase-type distributions: open problems and a few properties. Stoch. Model. **15**(4), 731–757 (1999)
133. Okamura, H., Dohi, T., Trivedi, K.S.: Markovian arrival process parameter estimation with group data. IEEE/ACM Trans. Netw. **17**(4), 1326–1339 (2009)
134. Okamura, H., Dohi, T., Trivedi, K.S.: A refined EM algorithm for PH distributions. Perform. Eval. **68**(10), 938–954 (2011)
135. Okamura, H., Dohi, T., Trivedi, K.S.: Improvement of expectation-maximization algorithm for phase-type distributions with grouped and truncated data. Appl. Stoch. Model. Bus. Ind. **29**(2), 141–156 (2012)
136. Olsson, M.: The EMpht-programme. Technical Report, Chalmers University of Technology (1998)
137. Osogami, T., Harchol-Balter, M.: A closed-form solution for mapping general distributions to minimal PH distributions. In: Kemper, P., Sanders, W.H. (eds.) Computer Performance Evaluation. Modelling Techniques and Tools. Lecture Notes in Computer Science, vol. 2794, pp. 200–217. Springer, Berlin (2003)
138. Osogami, T., Harchol-Balter, M.: Necessary and sufficient conditions for representing general distributions by Coxians. In: Kemper, P., Sanders, W.H. (eds.) Computer Performance Evaluation. Modelling Techniques and Tools. Lecture Notes in Computer Science, vol. 2794, pp. 182–199. Springer, Berlin (2003)
139. Panchenko, A., Thümmler, A.: Efficient phase-type fitting with aggregated traffic traces. Perform. Eval. **64**(7–8), 629–645 (2007)
140. Paxson, V., Floyd, S.: Wide area traffic: the failure of Poisson modeling. IEEE/ACM Trans. Netw. **3**(3), 226–244 (1995)
141. Rahnamay-Naeini, M., Pezoa, J.E., Azar, G., Ghani, N., Hayat, M.M.: Modeling stochastic correlated failures and their effects on network reliability. In: Proceedings of 20th International Conference on Computer Communications and Networks (ICCCN), pp. 1–6 (2011)
142. Realistic vehicular traces. http://www.lst.inf.ethz.ch/research/ad-hoc/car-traces/
143. Reinecke, P., Horváth, G.: Phase-type distributions for realistic modelling in discrete-event simulation. In: Proceedings of the 5th International ICST Conference on Simulation Tools and Techniques, SIMUTOOLS '12, pp. 283–290. ICST, Brussels (2012)
144. Reinecke, P., Krauß, T., Wolter, K.: Cluster-based fitting of phase-type distributions to empirical data. Comput. Math. Appl. **64**(12), 3840–3851 (2012)
145. Riska, A., Smirni, E.: ETAQA solutions for infinite Markov processes with repetitive structure. INFORMS J. Comput. **19**(2), 215–228 (2007)
146. Riska, A., Diev, V., Smirni, E.: An EM-based technique for approximating long-tailed data sets with PH distributions. Perform. Eval. **55**, 147–164 (2004)
147. Ruiz-Castro, J.E., Fernández-Villodre, G., Pérez-Ocón, R.: Discrete repairable systems with external and internal failures under phase-type distributions. IEEE Trans. Reliab. **58**(1), 41–52 (2009)
148. Sauer, C.H., Chandy, K.M.: Computer Systems Performance Modeling. Prentice Hall, Englewood Cliffs (1981)
149. Schmickler, L.: MEDA: mixed Erlang distributions as phase-type representations of empirical distribution functions. Stoch. Model. **8**(1), 131–156 (1992)
150. Schwarz, G.: Estimating the dimension of a model. Ann. Stat. **6**(2), 461–464 (1978)
151. Stewart, W.J.: Introduction to the Numerical Solution of Markov Chains. Princeton University Press, Princeton (1994)
152. Stewart, W.J.: Probability, Markov Chains, Queues, and Simulation. Princeton University Press, Princeton (2009)

153. Takahashi, Y.: Asymptotic exponentiality of the tail of the waiting-time distribution in a PH/PH/c queue. Adv. Appl. Probab. **13**(3), 619–630 (1981)
154. Telek, M., Heindl, A.: Matching moments for acyclic discrete and continuous phase-type distributions of second order. Int. J. Simulat. Syst. Sci. Tech. **3**(3–4), 47–57 (2002). [Special Issue on: Analytical and Stochastic Modelling Techniques]
155. Telek, M., Horváth, G.: A minimal representation of Markov arrival processes and a moments matching method. Perform. Eval. **64**(9–12), 1153–1168 (2007)
156. Thümmler, A., Buchholz, P., Telek, M.: A novel approach for phase-type fitting with the EM algorithm. IEEE Trans. Dep. Sec. Comput. **3**(3), 245–258 (2006)
157. Trivedi, K.S.: Probability and Statistics with Reliability, Queuing and Computer Science Applications, 2nd edn. Wiley, Chichester (2002)
158. Van Houdt, B., Lenin, R.B., Blondia, C.: Delay distribution of (im)patient customers in a discrete time D-MAP/PH/1 queue with age-dependent service times. Queueing Syst. **45**(1), 59–73 (2003)
159. Vehicular mobility trace of the city of Cologne, Germany. http://kolntrace.project.citi-lab.fr/
160. Wu, C.F.J.: On the convergence properties of the EM algorithm. Ann. Stat. **11**(1), 95–103 (1983)

Index

P. Buchholz et al., *Input Modeling with Phase-Type Distributions and Markov Models: Theory and Applications*, SpringerBriefs in Mathematics,
DOI 10.1007/978-3-319-06674-5,

U

75692003R00080

Made in the USA
San Bernardino, CA
03 May 2018